SIMON WILLARD

AND HIS CLOCKS

SIMON WILLARD

FROM A PORTRAIT
IN THE POSSESSION OF THE MISSES BIRD
DORCHESTER, MASS.

SIMON WILLARD

AND HIS CLOCKS

Formerly titled "A History of Simon Willard, Inventor and Clockmaker."
Together with some account of his sons—his apprentices—and
the workmen associated with him, with brief notices
of other clockmakers of the family name.

BY HIS GREAT GRANDSON

JOHN WARE WILLARD

DOVER PUBLICATIONS, INC.
NEW YORK

This Dover edition, first published in 1968, is an unabridged and corrected republication of the work originally printed in a limited edition of five hundred copies by E. O. Cockayne, Boston, in 1911 under the title *A History of Simon Willard, Inventor and Clockmaker.*

Standard Book Number: 486-21943-7
Library of Congress Catalog Card Number: 68-23802

Manufactured in the United States of America
Dover Publications, Inc.
180 Varick Street
New York, N. Y. 10014

TABLE OF CONTENTS

LIST OF ILLUSTRATIONS

_____

(Plates 21 and 22 are also reproduced in color on the inside covers.)

SIMON WILLARD

PREFACE

This book was originally intended simply as a memoir of Simon[5] Willard, Clock-maker and Inventor. While writing it the author found so many erroneous ideas prevailing about Simon[5] Willard, and his clocks, that the book was enlarged to its present form. The compilation of the book has been one of great difficulty. The material at hand is very scanty, and much of the early life of the Willards will have to remain blank for want of definite information. The author is particularly indebted to his father, Zabdiel[7] Adams Willard, for much of the information, particularly about the clocks, and the methods of manufacture, given in this book; indeed without his assistance it could not have been written. To Mr. Edwin A. Howe, Town Clerk of Grafton, Mass., the author expresses his thanks for his kindness in allowing him to examine the town records, and also to those who so kindly allowed him to take photographs of the clocks, portraits, documents, etc., used in illustrating this book, and without which the work would be of little interest.

The author cannot hope to have avoided all mistakes, but trusts there are no serious ones.

ABBREVIATIONS

O. G. R.	Original Grafton Records
R. C.	Boston Record Commissioner's Reports
B. M. & D.	Births, Marriages and Deaths
G. S.	Grave Stones

SIMON WILLARD

Simon[5] Willard (Benjamin[4] Joseph[3] Benjamin[2] Simon[1]) the celebrated clock-maker and inventor, a lineal descendant of the first New England progenitor of the family, Major Simon Willard, the founder of Concord, Mass., and prominent leader in King Philip's war, was one of a family of twelve children, born in Grafton, Mass., the eighth son of Benjamin[4] and Sarah[5] (Brooks) (Ebenezer[4] Noah[3] Joshua[2] Thomas[1]) Willard.[2]

Children of Benjamin and Sarah (Brooks) Willard.

I   Sarah, born July 7, 1740;[3] died Nov. 5, 1751.[5]
II   Joseph, born Dec. 27, 1741;[3] minister, H. C. 1765.[6]
III   Benjamin, born Mar. 19, 1743;[3] clock-maker.
IV   Solomon, born Jan. 8, 1745-6;[3] tanner.
V   Samuel, born Aug. 19, 1746;[3] died October 31, 1751.[5]
VI   John born Aug. 8, 1748;[3] lastmaker.
VII   Joshua, born May 18, 1751;[3] blacksmith.
VIII   Simon, born Apr. 3, 1753;[3] clock-maker.
IX   Ephraim, born Mar. 18, 1755;[3] clock-maker.
X   Aaron, born Oct. 13, 1757;[3] clock-maker.
XI   Lucy, born Oct. 10, 1759.[3]
XII   Eunice, baptized June 20, 1766.[4]

As the date of Simon Willard's birth has often been erroneously given, a photographic copy of the original Grafton Record (Plate 1) is shown. Very curiously the entire family with one exception is recorded on one page. The history of

[1]Willard Memoir. Pp. 365-383-433.
[2]Concord B. M. and D. Page 145.
[3]Original Grafton Records. Vol. 1. Page 206.
[4]Grafton Vital Records. Page 148.
[5]Original Grafton Records. Deaths. Vol. 1. Page 267.
[6]Willard Memoir. Page 433.

Simon Willard's early boyhood in Grafton is very meager. He had a limited schooling, in which the study of Latin figured somewhat; he attended the school of his native town. He did not take kindly to hard study, however, and showed such an inclination for mechanical pursuits, that at the early age of twelve his father apprenticed him to a Mr. Morris, "an Englishman then engaged in the manufacture of clocks in Grafton."[7]   Drake[8] also says that Simon Willard learned his trade of an Englishman named Morris. The author is obliged to confess after an exhaustive search that he is utterly unable to locate or identify this Morris. There are none of that name to be found in the early Grafton records, and there is nothing in any of the Worcester County Registry or Probate Records that give the slightest clue, nor in any of the surrounding towns is there any person, at that period, by the name of Morris, whose occupation is given as a clock-maker that the author can find. Histories of Grafton throw no light on the subject.   Drake gives no authority for his statement, nor Holden. If this Morris had an existence he probably was a Journeyman clock-maker, or a person that only had a rudimentary knowledge of clock-making. On the other hand, Simon Willard himself was heard to say that the man to whom he was apprenticed knew little or nothing of the art himself, and that his teacher was his brother Benjamin. Until more definite information is available, Simon Willard's statement will have to be accepted as the correct one. Benjamin Willard never made a very good clock, and possibly Benjamin learned the trade of this Morris, and Holden and

[7]Edward Holden. *Boston Transcript.* Sept. 4, 1857.
[8]R. C. Vol. 34. Page 152.

PLATE 1

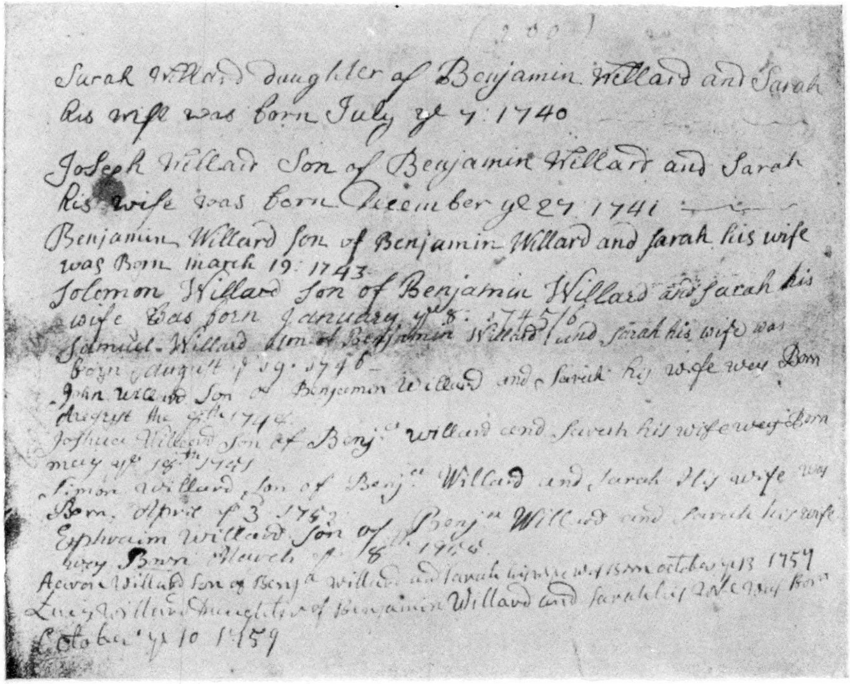
Sarah Willard daughter of Benjamin Willard and Sarah
his wife was born July ye 7: 1740
Joseph Willard Son of Benjamin Willard and Sarah
his wife was born December ye 27 1741
Benjamin Willard son of Benjamin Willard and sarah his wife
was Born march 19: 1743
Solomon Willard son of Benjamin Willard and sarah his
wife was born January ye 8: 1745/6
Samuel Willard son of Benjamin Willard and sarah his wife was
born August ye 19: 1746
John Willard Son of Benjamin Willard and Sarah his wife was Born
August the 9th 1748
Joshua Willard son of Benjn willard and sarah his wife was Born
may ye 18th 1751
Simon Willard son of Benjn Willard and sarah his wife was
Born April ye 3 1753
Ephraim Willard son of Benjn Willard and sarah his wife
was Born March ye 18th 1755
Aaron Willard son of Benja willard and sarah his wife was Born october ye 13 1757
Lucy Willard daughter of Benjamin Willard and sarah his wife was born
October ye 10 1759

PHOTOGRAPHIC REPRODUCTION OF THE ORIGINAL
GRAFTON RECORDS
OF THE
BENJAMIN WILLARD FAMILY

PLATE 2

HOUSE AND WORKSHOP OF SIMON WILLARD. WASHINGTON STREET. ROXBURY. MASS.
SIMON WILLARD'S HOUSE ON LEFT. MR. CHILD'S ON THE RIGHT

Drake have confused him as Simon's teacher. Whoever was
his instructor, Simon Willard on being apprenticed at once
found himself in his natural element, and so early did his
genius express itself that before the year was out he had
made with his own hands, without assistance from his master,
a clock that was at once pronounced far superior to those
produced by his master. This clock was the tall, striking
clock in general use up to the beginning of the nineteenth
century. When it is considered that there were no lathes, or
wheel cutting, or pinion shaping machines, in those days, and
that all work on a clock had to be done by hand, the file, the
drill, and the hammer being the only instruments, the feat of
making a clock at his age may be considered extraordinary.
Simon Willard was at this time but thirteen years of age.
From this time on, all his clocks were made by hand. How
long Simon Willard remained with his instructor, the author
has no means of knowing. All information in regard to his
early life, and that of his brothers, while in Grafton is vague
and unsatisfactory. If he finished his apprentiship, it is possible he might have worked for his brother Benjamin, who had
his clock factory there, but there is more reason to think, however, that he set up in business for himself, as clocks marked
"Simon Willard, Grafton," are occasionally found. The first
really authentic record we have of Simon Willard is during
the Lexington alarm, when with his brothers, John, Ephraim,
and Joshua, he marched with Capt. Aaron Kimball's company
of militia to Roxbury. His record is as follows:[9] — "Simon
Willard, Private, Capt. Aaron Kimball's co of militia [Col]
Artemas Ward's regt, which marched in response to the alarm

[9]Mass. Soldiers and Sailors of the Revolution. Vol. 17. Page 395.

of April 19, 1775, said Willard marched April 19, 1775, dis-
charged April 24, 1775. service 1 week. reported returned
home." Simon Willard was not warlike. After his discharge
he returned to Grafton and staid there during the war, he
understood making clocks, not war. He was often heard to
say that the musket he carried had no lock on it. He was
drafted into the army later on, but having no longing for a
military life, he procured a substitute. The substitute said he
preferred the cavalry to the infantry service, and so Simon
gave him his own horse, and provided accoutrements, and took
the man to the recruiting officer, who looked him over,
accepted him, and put his name on the roster. After the
ceremonies were completed, the recruit mounted his horse,
and rode off, after solemnly promising to return said horse
and accoutrements at the end of his enlistment. He was never
heard of again, probably having as little desire for military
glory as Willard himself. When asked about it, Simon
Willard's reply always was, "I suppose he is riding yet." In
fact, to tell the honest truth, Simon Willard had a mortal
dread of fire-arms. On one occasion, he was in the office
of his son-in-law, and seeing an old musket, took it up, and
in some way during his examination of the mechanism, it
went off with a tremendous report, sending a charge through
the ceiling and sending him sprawling on his back on the
floor. After this he never would touch a gun, and if he was
told the gun was not loaded his reply was always the same,
"Well it may go off if it isn't." Just how Simon Willard
was occupied during the period between 1775 and 1780, the
author is unable to say with any certainty, but it seems to be
reasonably certain that he was in the clock business for him-

self, for, as before noted, clocks marked "Simon Willard, Grafton," are sometimes found. He would have hardly marked them unless he had a shop of his own. During this period, he was married. Simon Willard married,[10] Nov. 29, 1776, Hannah Willard. She was born April 9, 1756,[11] died, Aug. 8, 1777.[12]

Children of Simon and Hannah Willard.

Issac Watts, born Feb. 6, 1777.[13]
died Aug. 8, 1777.[12]

Hannah[5] Willard (Joseph[4] Joseph[3] Benjamin[2] Simon[1]) was Simon Willard's first cousin.[14] She and her infant son died the same day from some epidemic prevailing in Grafton at that time. After the death of his wife, Simon Willard probably remained for a time in Grafton, Clock-making and perhaps peddling them around the country. Just when he left Grafton and came to Roxbury is uncertain. Drake[15] says he came in 1780, but gives no authority. In a vote of thanks given to Simon Willard, Aug. 20, 1829, by the Corporation of Harvard College, they say that for over fifty years he had charge of the clocks of the College. This certainly would show that Simon Willard was in Roxbury before 1780. He probably came sometime between 1777 and 1780. On his arrival in Roxbury, Simon Willard set up his shop in the building now numbered 2196 Washington St. (Plate 2) and which he occu-

---

[10]O. G. R. Marriages. Vol. 3. Page 94.
[11]O. G. R. Births. Vol. 1. Page 220.
[12]Family Record.
[13]O. G. R. Births and Deaths. Vol. 1. Page 104.
[14]History of Grafton, by F. C. Pierce. Pages 603, 604, 605.
[15]R. C. Vol. 34. Page 152.

pied until his retirement in 1839, a period of over fifty-eight years. The author is of the opinion that for the first few years Simon Willard did not live continuously at Roxbury, but went back and forth, perhaps spending his winters at Grafton, for in a deed dated March 2, 1778, recorded April 1785,[16] we find him buying an estate in Grafton of Nathan Morse, in the deed he calls himself "clock and watch maker of Grafton." He probably sold his estate at some later date, but the author has never been able to find any record of such a sale, probably never being recorded.

Simon Willard's name first appears in the Roxbury Records in 1783, when he is taxed for 2 polls, 3£ Real Estate, 6£ Personal.[17] Very curiously, his brother Aaron appears at the same date, taxed for exactly the same amount. As Simon Willard was taxed for two polls, it would seem to show that he had a workman, or an apprentice with him. The census of 1790[18] shows him to be a resident of Roxbury, but his name is misspelled Williard. His name, however, is not in the index. The house (Plate 2) in which he did all his work, made all his inventions, set up all his clocks, and where all his children were born, is still standing in Washington St. (formerly Roxbury St.), with very little alteration, and with the exception of the show windows, cut in what was then the end of the house, and the removal of the pitched roof, is (1905) practically as Simon Willard left it. The first floor contained a parlor, back of which was the kitchen with a huge open fire-place. A very fine Hall Chime clock stood in this kitchen next to the

[16]Worcester Deeds. Vol. 97. Page 504.
[17]Roxbury Tax Lists for November, 1783.
[18]First Census of United States for Massachusetts, 1790. Roxbury Town. Page 205.

entrance. It would be of interest to know what became of this clock. The sleeping rooms, four in number, were above. How his family of eleven children found room in this house is a matter of astonishment to the present generation, who do not understand the possibilities of the trundle bed, a convenient institution in those days. The entrance to the house was from the side. Back of the kitchen was the work-shop, and under it was a shed in which was stored the family wood and provisions. In the floor of the shop was a hole about a foot square through which went the pendulum of his great Turret clocks, when tested for time. The principal part of the pendulum swung below the floor in the centre of the shed, and thereby hangs a tale. In the early part of the century, half a dozen pirates, more or less, were executed on Boston[19] Neck, now Washington St., about half-way between Roxbury and Boston. Of course, the hanging was made a gala occasion, the whole population turned out en masse to see the show. Among these were half a dozen Roxbury School boys, who were so elated by the exhibition that they concluded to have a private hanging of their own, and so adjourned to Simon Willard's shed to carry out the idea with all the particulars. One of their number, Johnty Collins, so named, not over bright, was selected as the criminal, accused of piracy, with un-numbered atrocities, tried, convicted and ordered to instant execution. The desperate villain was placed on a nail keg, a noose adjusted round his neck, secured to a convenient floor beam, and drawn taut. The nail keg was kicked away, and Johnty was left dangling in mid air, making desperate efforts to get away. As he was growing black in the face,

[19]R. C. Vol. 34. Page 68.

the court jury fled precipitately, making no effort to take him down, and he would undoubtedly have perished then and there had not Mr. Willard hearing some unearthly gasping and other unaccustomed sounds coming up through the pendulum hole, went down to investigate and just in time to prevent the unhappy Johnty from sharing the fate of the other pirates. The author would say here, that when he visited Simon Willard's old house with his father, Z. A. Willard, in 1905, it was practically as described. At a later date, 1909, the author had a conversation with the owner of the building, Mr. Benj. F. James who informed him that the house was remodelled in 1861, the pitched roof taken off and a flat one substituted and other alterations made. If this was so, the house must have been made over on the original lines, for Z. A. Willard said (1905) that as far as he could see the house was nearly as he knew it in 1838, the rooms very little altered, the doorway looked the same, and he especially pointed out to the author the shed and the pendulum hole, and told the story of the hanging. The present occupant of the place has pulled it about so that very little of the interior can be recognized. Shortly after Simon Willard's arrival in Roxbury, the exact date not being certain, he made a large double dial clock, for the purpose of advertising his business. His own house not being strong enough to stand the weight of this clock, he put it up on the front of his next door neighbor's, Mr. Child's house. Here it remained, a land-mark for this part of Roxbury. After Simon Willard retired from business, in 1839, he presented this clock to the town of Roxbury, and it remained for many years after, until it was finally taken down. The staples that held the clock are still to be seen, between the upper windows, and are plainly shown in the picture (Plate 2), the bay window is an addition of recent years. The author quotes an extract from a letter received by him in regard to this clock.[20] "I am sorry I cannot remember the name of the private collector to whom I disposed of the old clock, several years ago. The old clock laid in the cellar for years until this person I write of heard of it as being a land-mark, looked it up, and as we set no value on it, were glad to have him give it a place with his collection of antiques." The author regrets he has been unable to locate the collector in order that a photograph might have been obtained of the old clock. Whatever intentions Simon Willard had of returning to Grafton were changed by his marriage in 1788, and he became a permanent resident of Roxbury.

Simon[5] Willard married, 2nd, Jan. 23, 1788,[21] Mrs. Mary (Bird) Leeds, widow of Richard Leeds of Dorchester,[22] Mass., and daughter of Edward and Mary (Star) Bird.[23] She was born Feb. 18, 1763,[24] died July 23, 1823.[25]

Children of Simon and Mary (Bird) Willard.

1. Thomas Rice, born Roxbury, Nov. 3, 1788.[26]
2. Hannah, born Roxbury, March 25, 1790.[26]
3. Harriot, born Roxbury, Sept. 26, 1791.[26]
4. Mary, born Roxbury, March 12, 1793.[26]

[20]Letter from Mr. Benjamin F. James, Roxbury. January 28, 1908.
[21]Roxbury Marriages, 1632 to 1860.
[22]R. C. Vol. 21. Page 235.
[23]R. C. Vol. 28. Pages 289 and 344.
[24]R. C. Vol. 21. Page 164.
[25]Grave Stone, Forest Hills Cemetery.
[26]Family Records.

5. Simon, born Roxbury, Jan. 13, 1795.[26]
6. Joseph, born Roxbury, Sept. 13, 1796.[26]
7. Julia Knox, born Roxbury, Sept. 25, 1798.[26]
8. John Mears, born Roxbury, March 20, 1800.[26]
9. Julia, born Roxbury, Jan. 28, 1802.[26]
10. Benjamin Franklin, born Roxbury, Nov. 2, 1803.[26]
11. Sarah Brooks, born Roxbury, June 25, 1805.[26]

A more extended notice of some of the children of Simon Willard will be given in other chapters, and it will be noticed how his mechanical ability and inventive faculties were inherited and carried through another generation. Simon Willard had not been in Roxbury long before his inventive faculties asserted themselves, and he brought out his Clock Jack, for which, in 1784, he was granted the exclusive privilege of making and selling, by an act of the General Court of Massachusetts, passed July 2, and approved by John Hancock.[27]

1784.   CHAPTER 17.
MAY SESSION. CH. 17.

"An Act Granting to SIMON WILLARD, The Exclusive Privilege of Making and vending CLOCK JACKS, FOR FIVE YEARS. Chap. 17.

Whereas it appears that it will be productive of great national advantages that every reasonable encouragement should be given to arts, science, useful inventions, and improvements. And whereas, Simon Willard of Roxbury, hath by study and application, invented a clock jack with a compleat apparatus, which appears well calculated to answer the end designed and hath petitioned this Court for an exclusive patent for making and vending the same.

Be it therefore enacted by the Senate and House of Representatives in General Court assembled, and by the authority of the same, That there be granted unto the said Simon Willard, the sole and exclusive right to make and sell his said clock jacks, within the Commonwealth for and during the term of five years next ensuing.

[26]Family Records.
[27]Laws and Resolves of Massachusetts, 1784. Page 45.

PLATE 3

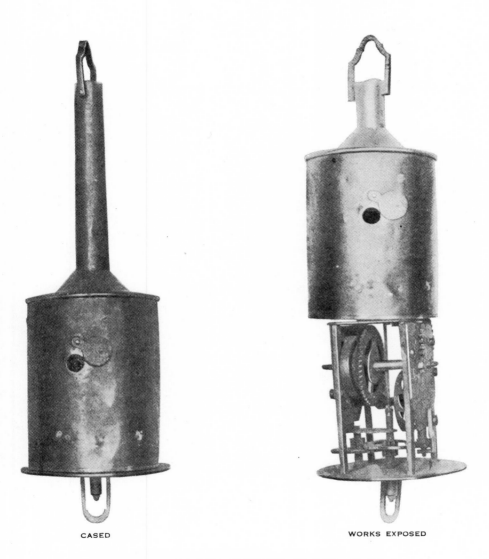

CASED   WORKS EXPOSED

SIMON WILLARD PATENT CLOCK JACK

And be it further enacted by the authority aforesaid That no person shall, from and after the passing of this act, and during the said term of five years, make, sell or utter clock jacks in imitation of those invented by the said Willard, without his licence and approbation. And be it further enacted by the authority aforesaid That if any person, shall, from and after the passing of this Act, and during the said term of five years, make, sell or utter clock jacks, as aforesaid, he, she, or they, so offending, shall, for every such offense forfeit and pay the sum of six pounds, one moiety thereof to the use of this Commonwealth and the other moiety to the person, who shall sue for the same to be recovered in an action of debt, in any Court proper to try the same.

Provided always and be it enacted by the authority aforesaid, That the said Jacks shall at no period during the said term of five years be sold by the said Willard for a greater sum than three pounds, and the said exclusive right granted in manner as aforesaid, shall cease and determine immediately upon the said jacks being raised by the said Willard to a greater sum."

July 2, 1784.

For the benefit of the uninitiated, a Clock Jack may be described as a piece of kitchen furniture much used in old times for roasting meat. The utensil was suspended by a hook from the mantel shelf in front of the open fire place and the meat was hung on the hook at the end of the chain, and the machinery being wound up, the meat was slowly rotated. These clock jacks, mostly imported from England, were very heavy and cumbersome. Simon Willard's improvement consisted in making the whole instrument lighter and more compact and having the machinery actuated by a spring and lever, on the principle of the verge escapement, and was perhaps suggested by his watch. The whole was enclosed in a neat brass case (see Plate 3). Simon Willard never made very many of these clock jacks, for about the same time somebody brought out the tin kitchen, which proved more convenient as long as the open fire place was used for cooking.

After leaving Grafton, Simon Willard abandoned the manufacture of the Half or Shelf clock, and devoted himself exclusively to the making of the Hall clock, Church or Turret clocks, Gallery clocks, and general repair work. In the summer time, or when business was slack, he peddled clocks about the country, his beat was along the North Shore. Simon Willard never advertised in the papers, the nearest approach to an advertisement is the printed form (Plate 4) that is sometimes, though rarely, found inside the doors of his Hall clocks. It will be noticed that the forms were printed in Worcester, Mass., but no date is to be found on them. At the top of the form will be noticed the picture of his Dial clock, before spoken of. The perambulators referred to in his advertisement were the forerunners of the modern odometer.

CLOCK MANUFACTORY.
SIMON WILLARD,

AT his Clock Dial, in Roxbury Street, manufactures every kind of CLOCK WORK; fuch as large Clocks for Steeples, made in the beft manner and warranted, price with one dial, 500 dollars; with two dials, 600 dollars; with three dials, 700 dollars with four dials, 900 dollars. — Common eight day Clocks with very elegant faces and mahogany cafes, price from 50 to 60 dollars.—Elegant eight day Time pieces, price 30 dollars.—Time pieces which run 30 hours, and warranted, price 10 dollars. Spring Clocks of all kinds, price from 50 to 60 dollars.—Clocks that will run one year, with once winding up, with very elegant cafes, price 100 dollars.— Time pieces for Aftronomical purpofes, price 70 dollars.—Time pieces for meeting houfes, to place before the gallery with neat enamelled dials, price 55 dollars. Chime clocks that will play 6 tunes, price 120 dollars.— Perambulators are alfo made at faid place, which can be affixed to any kind of wheel carriage, and will tell the miles and rods exact, price 15 dollars.

GENTLEMEN who wifh to purchafe any kind of CLOCKS, are invited to call at faid WILLARD'S CLOCK MANUFACTORY, where they will receive

PLATE 4

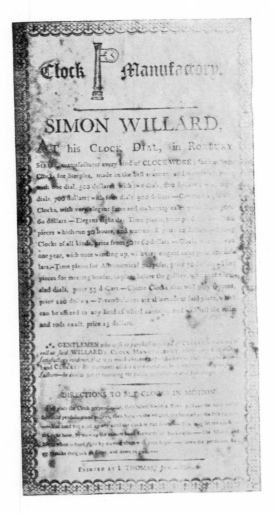

SIMON WILLARD

CLOCK ADVERTISEMENT

fatisfactory evidence, that it is much cheaper to purchafe new, than old and fecond hand CLOCKS: He warrants all his work — and as he is ambitious to give fatis-faction — he doubts not of receiving the public approbation and patronage.

DIRECTIONS TO SET CLOCKS IN MOTION.

Firft place the clock perpendicular, then faften it with a fcrew, pull out the nails which faften the pendulum and pulleys, then hang on the weights, the heavieft on the ftriking part.—You need not wind up any until the clock is run down.—You may fet the clock to the right hour, by moving the minute hand forwards or back-wards.—The Month and Moon wheel is fixed right by moving them with your finger.—Screw the pendulum ball up to make the clock go fafter, and down to go flower.

PRINTED BY I. Thomas, Jun.—Worcefter.

In 1801, he invented an improved Timepiece, and applied for a patent for it. This patent was granted and issued to him by the U. S. Patent Office, February 8, 1802. This Willard Patent Timepiece was a great improvement on the tall upright clock then in use. It was smaller, more compact, more easily set up, and handled, and being made to fasten to the wall, was not always getting knocked off and getting smashed like the Half or Shelf clocks, and were much cheaper than the old style clocks. It was an instant and complete success, coming at once into public favor, and superseded all other clocks. It was a perfect timekeeper, and beautifully simple in construction. A curious fact of this invention is that although hundreds of thousands of these clocks have been made to the present day, not the slightest improvement has been made upon them by any of the ingenious workmen who have made the clocks up to the present time. This fact would place him in the front rank of the Horological artists, with John Harrison (1693-1776), inventor of the Marine Chronometer, and yet Harrison's invention was greatly improved by John Arnold (1736-1799), who invented the Detent Escapement, and Thomas Earnshaw (1749-1829), who made further improvements, making it the most perfect instrument for determining longitude now in use. Another thing that served to make this Timepiece popular was its graceful shape. The original patent issued (Plate 5), signed by President Thomas Jefferson, James Maddison, Secretary of State, and Levi Lincoln, Attorney-General, is still preserved in good condition.

"The Schedule referred to in these Letters Patent, and making part of the same) (containing a description in the words of the said Simon Willard himself of his im-) (provement, In a Timepiece.

Be it known, that I, Simon Willard of Roxbury, in the County of Norfolk and Com-) (monwealth of Mafsachusetts, have invented, constructed and applied to use a new and useful) (regulator, or timepiece. The description of its machinery, and the explanation of its principles are) (as follows, viz:— The height of it is two feet, and the diameter of the face about seven inches) (but may be increased or diminished to any size. The power of motion is a weight instead) (of a spring, which is the case in all regulators, and time pieces smaller than the clock. The) (weight falls only fifteen inches in eight days, during which time the regulator goes without) (winding up, whereas, the weight of the eight day clock falls not less than six feet in the) (same time. By the construction of the timepiece or regulator the pendulum is brought forward) (in front of the weights, by which means it may be made longer and will consequently vibrate) (more accurately than the common method in which the pendulum was placed behind the) (weights. The pendulum is suspended on pivots by which it is prevented from the least warble) (which is one great cause of inaccuracy in the common regulator, considered hitherto irremidiable). (At the bottom of the pendulum there is a plate graduated to the arch described which serves to) (measure the oscillations of the pendulum and shews to what distance it ought to vibrate, from) (the center of gravity to keep true time and likewise shews when it is out of order or wants) (oiling or cleaning: also the variations of the vibratory motion of the pendulum from the influence) (of heat and cold, may be accurately ascertained and in a great measure remidied. The cace) (of the Regulator is thick glass, painted, varnished and gilt in a manner which can never) (fade, and is

PLATE 5

The United States of America.

To all to whom these Letters Patent shall come:

WHEREAS *Simon Willard a citizen of the State of Massachusetts* in the United States, hath alledged that he has invented a new and useful improvement

which improvement has not been known or used before his application; has *made oath* that he does verily believe that he is the true inventor or discoverer of the said improvement, *..* has paid into the Treasury of the United States, the sum of thirty dollars, delivered a receipt for the same, and presented a petition to the Secretary of State, signifying a desire of obtaining an exclusive property in the said improvement, and praying that a patent may be granted for that purpose: THESE ARE THEREFORE to grant, according to law, to the said *Simon Willard* his heirs, administrators, or assigns, for the term of fourteen years, from the *eighth* day of *the present month of February;* the full and exclusive right and liberty of making, constructing, using, and vending to others to be used, the said improvement, a description whereof is given in the words of the said *Simon Willard* himself, in the schedule hereto annexed, and is made a part of these presents:

IN TESTIMONY WHEREOF, *I have caused these Letters to be made Patent, and the Seal of the United States to be hereunto affixed.*
GIVEN *under my hand, at the City of Washington this eighth day of February in the year of our Lord, one thousand eight hundred & five and of the Independence of the United States of America, the twenty sixth.*

Th Jefferson

the PRESIDENT,

James Madison Secretary of State.

City of *Washington*

I DO HEREBY CERTIFY, That the foregoing Letters Patent, were delivered to me on the *eighth* day of *February* in the year of our Lord, one thousand eight hundred *and five* to be examined; that I have examined the same, and find them conformable to law. And I do hereby return the same to the Secretary of State, within fifteen days from the date aforesaid, to wit:—On this *eighth* day of *February* in the year aforesaid.

L. or Lincoln actg. Gov. U.S.

PATENT ISSUED TO SIMON WILLARD FOR A TIMEPIECE

IN POSSESSION OF THE AUTHOR

PLATE 6

SCHEDULE OF CLAIMS FOR PATENT GRANTED TO SIMON WILLARD FOR A TIMEPIECE

FROM ORIGINAL DOCUMENT

OF WHICH THE FIRST PAGE IS SHOWN ON PRECEDING PLATE

more durable as well as beautiful and cheaper than the common china ena-)
(melled or any other kind of caces. The door of the regulator is set with
glass painted and) (gilded with an oval space left through which the motion
of the pendulum is seen which has) (a pleasing effect. The whole of this regulator can be constructed and made with much less labor) (and expense
than any other kind of regulator yet constructed. In testimony that the
afore-) (mentioned is a true description thereof, I, the said Simon Willard have
hereunto set my) (Hand and Seal this twenty-fifth day of November in the
year of our Lord, one thousand,) (eight hundred and one.
Signed and Sealed in presence of us   Simon Willard.
    Luther Richardson.
    Thos. J. Robinson.

The specifications are very simple, not to say crude, and
were evidently made out by Simon Willard himself. They
would hardly pass muster in the Patent Office at the present
day, when every point is rigidly specified. There is much
uncertainty about the exact date when Simon Willard first
made his Timepiece, and there is good reason to think he
made them for some years before he patented them. As he
never dated his Timepieces it makes the question difficult.
Referring again to his advertisement (Plate 4) it will be
noted he mentions Timepieces. This advertisement was
printed by Isaiah Thomas, Jur., of Worcester, Mass., but unfortunately no date is given. Isaiah Thomas, Jur. was the son
of Isaiah Thomas, printer and publisher, of Boston, who
removed to Worcester at the outbreak of the Revolution,
and published the *Worcester Spy*, and other papers. Isaiah
Thomas, Jur. succeeded to the business in 1801. As some of
the clocks containing this advertisement are positively
known to have been made in 1790, it would seem to show
that the Timepiece was made at that date, supposing Isaiah
Thomas, Jur. was in business for himself at that time. There

is a tradition in the family that Simon Willard did not realize the value of his invention, and when he visited Washington to show the authorities how to run the clock he made for them (see letter, Page 18), President Jefferson saw the importance of the invention, and told Simon Willard to take out a patent for it. Until some documentary evidence is found giving an earlier date, the author is reluctantly compelled to give 1801 as the beginning of the Timepiece. Two letters (Plates 7 and 8) have recently (1909) come to light wherein his Timepieces are mentioned, and they give no earlier date than 1802. It would seem, however, from the data given about the Timepiece belonging to Mr. Dwight M. Prouty (Pages 49-50) that Simon Willard certainly was making the Timepiece as early as 1796.

In 1819 he applied for and obtained another patent, this time for an alarm clock. Part of the original Patent still exists (Plate 9). The cover is missing, evidently having been taken by some one, probably for the autographs. It would have had the signature of President Monroe. The wording of both the specifications is very quaint.

"The Schedule referred to in these Letters Patent, and making) (part of the fame containing a description in the words of the faid Simon Willard, himself) (of his improvement in Clocks.

There is an alarm wheel with teeth like the pallat wheel of a clock) (with fmall pallats: the hammer ftem being attached to the pallats, and reaching through) (the clock plates on the other side: and when let off, it ftrikes on the top of the cafe of) (the clock, and makes a noise like fome one rapping at the door, and it will wake you) (much quicker than to ftrike on a bell in the usual way.

There is only one wheel to the alarm part, with a little barrel which) (the ftring winds round a few times, and by pulling the little weight winds it

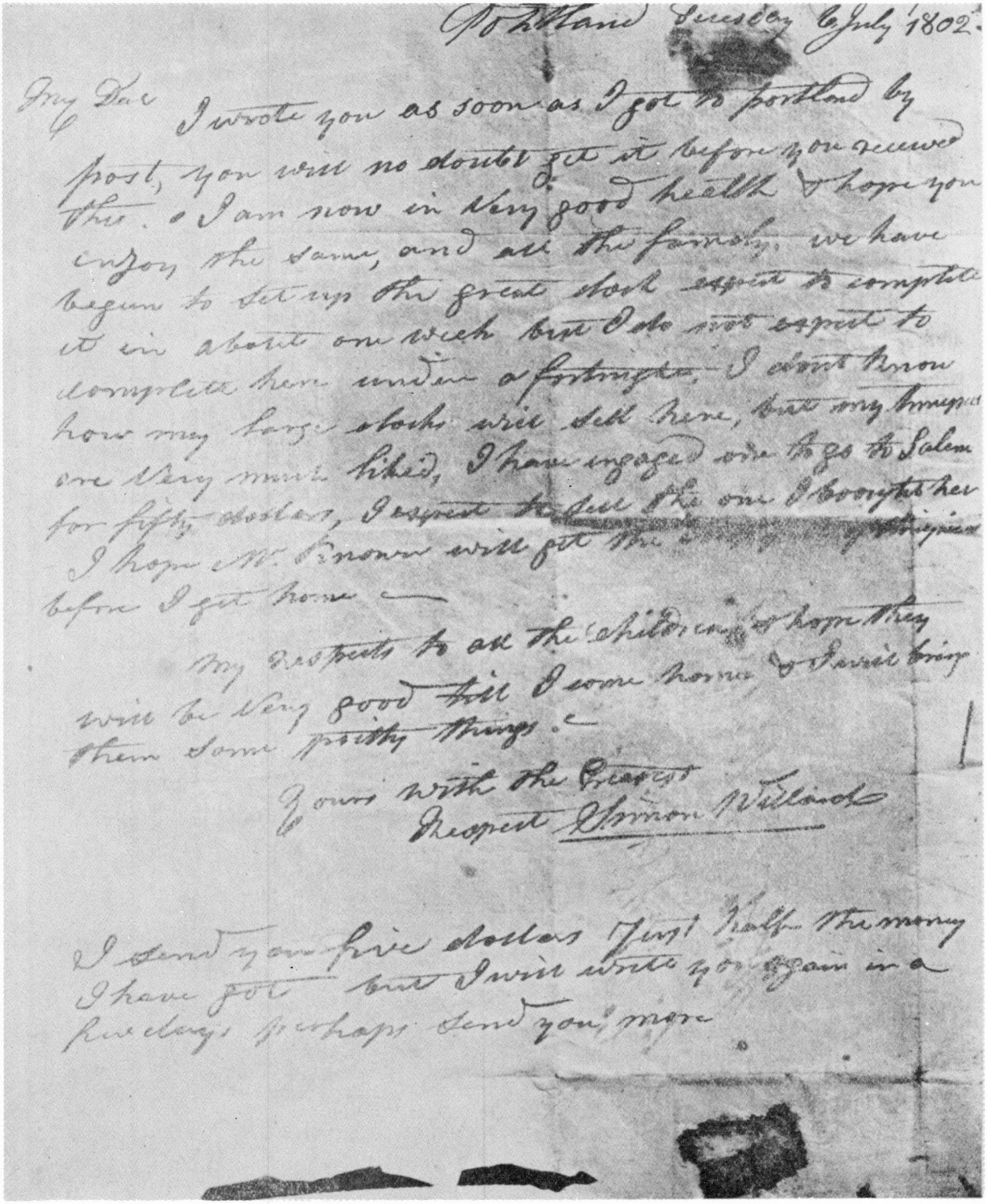

Portland Tuesday 6 July 1802

My Dear

I wrote you as soon as I got to portland by post, you will no doubt get it before you receive this. I am now in very good health & hope you enjoy the same, and all the family. we have begun to set up the great clock expect to complete it in about one week but I do not expect to complete here under a fortnight. I dont know how my large clocks will sell here, but my timepieces are very much liked, I have engaged one to go to Salem for fifty dollars, I expect to sell the one I brought here [illegible] I hope Mr. Brown will get the [illegible] before I get home —

My respects to all the children & hope they will be very good till I come home & I will bring them some pretty things —

Yours with the Greatest Respect Simon Willard

I send you five dollars Just half the mony I have got but I will write you again in a few days & perhaps send you more

LETTER FROM SIMON WILLARD TO HIS WIFE

PLATE 8

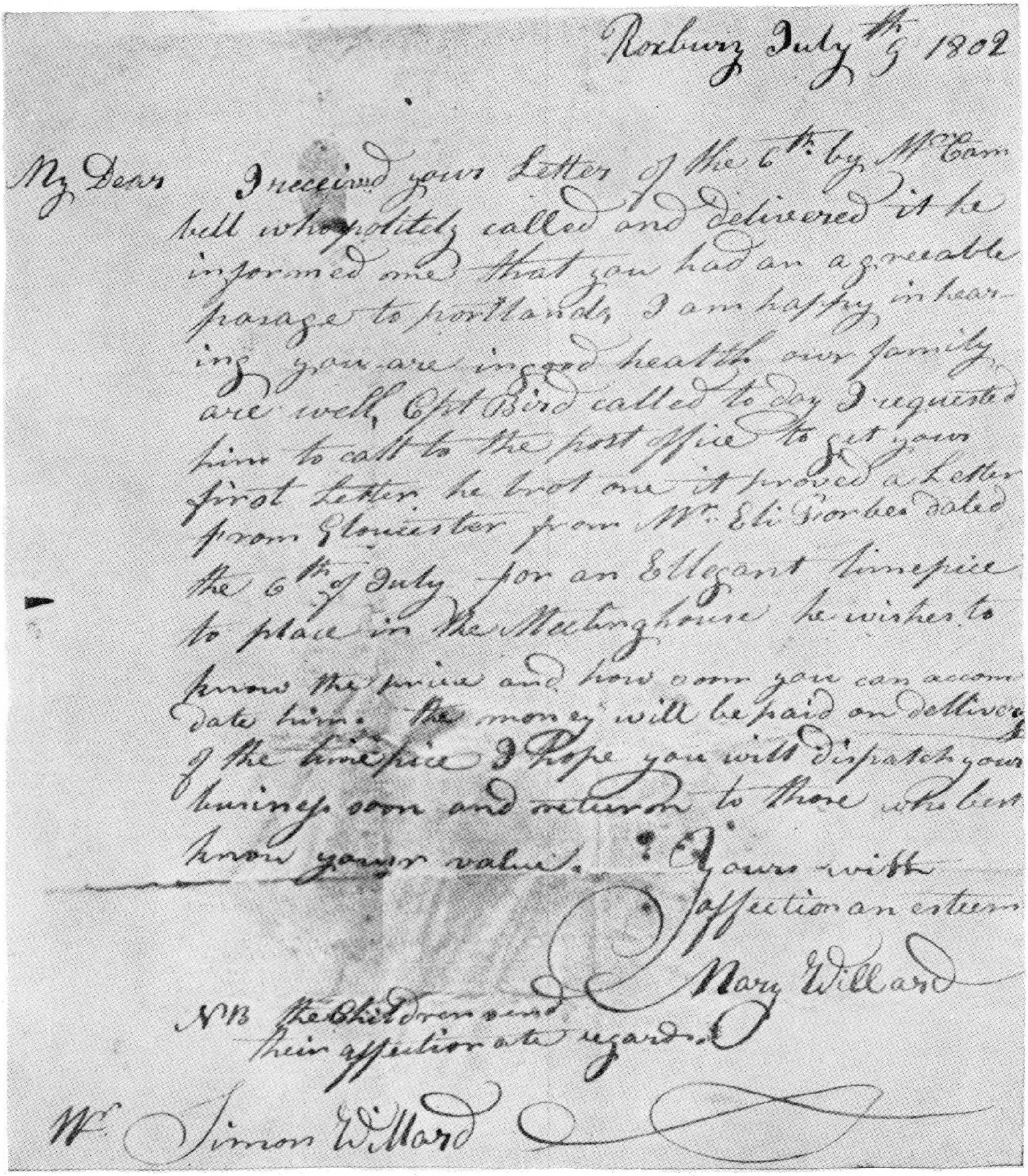

Roxbury July 9th 1802

My Dear I received your Letter of the 6th by Mr Cambell who politely called and delivered it he informed me that you had an agreeable passage to portlands I am happy in hearing you are in good health our family are well, Cpt Bird called to day I requested him to call to the post office to get your first Letter he brot one it proved a Letter from Gloucester from Mr. Eli Forbes dated the 6th of July for an Ellegant timepice to place in the Meetinghouse he wishes to know the price and how soon you can accomodate him. the money will be paid on delivery of the timepice I hope you will dispatch your business soon and return to those who best know your value. Yours with affection an esteem

Mary Willard

NB the Children send their affectionate regards

Mr Simon Willard

MRS. WILLARD TO HER HUSBAND

REPLY TO LETTER SHOWN ON PRECEDING PLATE 7

PLATE 9

The Schedule referred to in these Letters Patent and making part of the same, containing a description in the words of the said Simon Willard himself of his improvement in Clocks.

———————

There is an alarm wheel with teeth like the pallat wheel of a clock, with small pallats the hammer stem being attached to the pallats, and reaching through the clock plates on the other side, and when let off, it strikes on the top of the case of the clock, and makes a noise like some one rapping at the door, and will wake you much quicker than to strike on a bell in the usual way.

There is only one wheel to the alarm part, with a little barrel which the string winds round a few times, and by pulling the little weight winds it up. the little weight hangs outside of the case. You have no trouble in opening any part to wind it up.

The face of the clock is about 5 inches in diameter, with a little dial in the center, with twelve figures on it; by turning that dial you set it to the hour by a point in the hour hand, to the hour you wish to rise. The whole of the alarm part is entirely new, and very simple; and it is made upon a plan which will not fail in going off.

The time part, by having an intermediate wheel, will run eight days; and the weight only descends 13 inches in the eight days. The whole case is about 13 inches high, and easily moved to any part of the house, without putting it out of order.

The whole of the clock work is inclosed with a handsome glass, and it is wound up without taking the glass off, which prevents the dirt from getting into it. The whole plan of the clock I claim as my invention. The Pendulum is suspended upon & connected with the *Simon Willard*.

Witnesses
William Elliott
Robert Fenwick

SCHEDULE OF CLAIMS FOR PATENT GRANTED TO SIMON WILLARD FOR AN ALARM CLOCK
FROM ORIGINAL DOCUMENT IN THE POSSESSION OF THE AUTHOR

up, the) (little weight hangs outside of the cafe, you have no trouble in opening any part to) (wind it up.

The face of the clock is about 5 inches in diameter, with a little dial) (in the center, with twelve figures on it; by turning that dial you fet it to the hour) (by a point in the hour hand, to the hour you wish to rise.   The whole of the) (alarm part is entirely new, and very fimple, and it is made upon a plan which will not *fail* in going off.

The time part by having an intermediate wheel will run eight) (days, and the weight only descends 12 inches in the eight days.   The whole) (cafe is about 15 inches high, and easily moved to any part of the houfe without) (putting it out of order.

The whole of the clock work is inclosed with a handsome glafs) (and it is wound up without taking the glafs off, which prevents the dirt from) (getting into it.   The whole plan of the clock I claim as my invention.   The) (Pendulum is suspended upon and connected with the pivot.

      Witness.   Simon Willard.
      William Eliot.
      Robert Fenwick.

The author has never seen one of these alarm clocks made by Simon Willard, but all the other clock-makers of his time promptly copied it. He does not appear to have ever made very many of them, and they do not appear to have been a success, perhaps because people did not like to get up early in those days, any better than they do now. He also made the machinery for the early Revolving Lights of the sea coast. Some of his best work was done in making the Church or Turret clocks. In his repair work he altered and improved clocks of other makers, so that they were better than they were originally. After his Patent of 1802 was granted, he abandoned the manufacture of the Hall clocks almost entirely, and devoted himself to the Timepiece, only making the Hall clock on an order.

In 1801 he made a large clock for the United States

(copy)   Roxbury Jany 18th 1802 —

Dear Sir,

   I have received all your favors, the last
of which gives me disagreeable feelings as it respects
the Clock stoping, I think it must be owing to putting it up or to being transported such a distance.
I paid the greatest attention to have the Clock
well finished & regulated and well packed,
that it might be sure to go well; but I hope
you will not meet with much difficulty in
Setting it agoing — the clock is made upon a plan
not subject to stop or get out of order, it will I
think go twenty years without cleaning as the
movement part is made almost air tight; &
I think the clock will keep the most accurate
time. if it should gain time or loose a little
the person who takes care of it, can regulate
              it

t by screwing the pendulum up or down.

I hope it will give them satisfaction and not think the price much too high. I am making a clock for Portland meeting-house for which I am to have 750 dollars, and the expense and time I was at about the Senate clock, is as much as the Portland clock – clock-makers here who saw it did not think it too high. some gentlemen who saw it thought I ought to have more. however I shall be satisfied with a deduction, but I hope it will not be more than 100 dollars.

(signed) Simon Willard

To Sam: A. Otis

Senate at Washington. As this clock was made on the principle that Simon Willard afterwards patented, the authorities did not understand how to run it, and Simon Willard was obliged to go to Washington, and show them how to run it. While there he was introduced to Thomas Jefferson, then President, and the meeting developed into a very strong friendship between the two. The letter (Plate 6), explains itself; very curiously the bill for this clock has survived.

These two papers were found in the collection of the late Ben Perley Poore. This clock was destroyed when the British burned Washington in 1814. The author will observe here that no matter whether Simon Willard was making or losing on a clock, he put into it his very best in labor and material. In 1826, Simon Willard made a Turret clock for the University of Virginia, at Charlottesville, Va. It was ordered by Thomas Jefferson, who made out the plans and specifications, which were sent to Simon Willard, June 4, 1826. Simon Willard was often heard to say regarding these plans and specifications that they were the only ones he ever received while in business that were properly made out. The measurements were so accurately given and the plans so clearly drawn that when he put the clock up, everything fitted to the sixteenth of an inch. The correspondence about this clock is interesting, and is given below, being the original correspondence between Jefferson and Mr. Joseph Coolidge, Jr., Boston[28] (the husband of Ellen Wayles Randolph, Jefferson's granddaughter).

"There stands in the entrance of the Brooks Museum the first bell ever

[28]From the Alumni Bulletin of the University of Virginia. February, 1899. Pages 111-113.

United States of Am^a
to Simon Willard Dr.
to a clock for the Senate of United States at Washington — $750
to packing said clock — — — — — 20
$.770
Rec'd pay,
Roxbury Novr. 21. 1801

used in the University of Virginia, which was ordered by Mr. Jefferson in 1826 and which did good service for over sixty years.

Last April, on the 155th anniversary of Mr. Jefferson's birth, his great granddaughter, Miss Caroline Ramsey Randolph, granddaughter of his beloved daughter, Martha Jefferson Randolph, presented to the University of Virginia, through the Albemarle Chapter of the Daughters of the American Revolution, the original correspondence between Mr. Jefferson and Mr. Coolidge of Boston (the husband of Ellen Wayles Randolph, Mr. Jefferson's granddaughter), regarding the purchase of this bell and the clock which remained in use until its destruction in the fire of 1895. These valuable papers will shortly be placed near the bell, and together they will form interesting illustrations of this great man's forethought and care for the University in the very last days of his life.

The first of these letters bears the date " Monticello, April 12, 1825," and Mr. Jefferson writes thus after speaking of the opening of the University: " Your kind disposition towards our University will sometimes I fear be the source of trouble to you. We understand that the art of bell-making is carried to greater perfection in Boston than elsewhere in the United States. We want a bell which can *generally* be heard at the distance of two miles, because

this will ensure its being *always* heard at Charlottesville. As we wish it to be sufficient for this, so we wish it not more so, because it will add to its weight, price, and difficulty of management. Will you be so good as to enquire what would be the weight and price of such a bell and inform me of it?
———— ————"

The second of these letters by Mr. Jefferson written one year later explains the reasons for the delay in giving the final order, and is as follows:

"Monticello, June 4, 1826.

DEAR SIR. You have heretofore known that the ability of the University to meet the necessary expense of a bell and clock depended on the remission by Congress of the duties on the marble bases and capitals used in our buildings, a sum of nearly $3,000. The remission is granted, and I am now authorized to close with Mr. Willard for the undertaking of the clock as proposed in your letter of August 25. I must still, however, ask your friendly intermediacy because it will so much abridge the labors of the written correspondence, for there will be many minutiæ which your discretion can direct, in which we have full confidence and shall confirm as if pre-directed. I have drawn up the material instructions on separate papers, which put into Mr. Willard's hands will, I trust, leave little other trouble for you. We must avail ourselves of his offer (expressed in same letter) to come himself and set it up, allowing the compensation which I am sure he will make reasonable. The dial plate had better be made at Boston, as we can prepare our aperture for it of sixty inches with entire accuracy. We wish him to proceed with all practicable dispatch, and are ready to make him whatever advance he usually requires, and we would rather make it immediately, as we have a sum of money in Boston which it would be more convenient to place in his hands at once, than to draw it here and have to remit it again to Boston. If it would be out of his line to engage for the bell also, be so good as to put it into any hands you please, and to say what we should advance for that also.———— ————"

The instructions referred to by Mr. Jefferson in the preceding letter were as follows:

"INSTRUCTIONS FOR THE GOVERNMENT OF THE ARTIST IN MAKING THE CLOCK FOR THE UNIVERSITY OF VIRGINIA.

"The bell is to weigh 400 lbs., which it is supposed will insure its being heard 1½ miles under any circumstances of weather. The distance of the hollow cylinder in which the weights are to descend, and its oblique direction from the dial plate has rendered necessary an outline of the ground plat and elevation of the parts of the building where it will be placed; this is drawn on lined paper, in which every line counts a foot, and every 10th line is more strongly drawn to facilitate counting, by which the measures are to be taken and not all by scale and compass. The cylindrical space in which the weights descend is of 5 ft. diameter and 48 ft. depth, that is to say from the level of the center of the dial plate to the ground. The tympanum of the pediment, in the center of which the dial-plate is to be placed, is 42 ft. in the span, and 9 ft. 4 in. in its perpendicular at the apex, that is to say the naked of the tympanum within its cornice. Such a triangle admits a circle of 52 in. radius to be inscribed within it, so that describing in its center the dialplate of 30 in. radius, and around that the architrave 10 in. wide, there will remain a clear space between the architrave and cornice of the pediment of 12 in. in the points where they approach nearest. But the dial-plate must be as much wider than the 5 ft. which it shows as to fill a rabbet of ⅓ in. at least in the back face of the architrave in which it may be firmly imbedded. It must be of metal of course, as wood would go too soon into decay.

"The face of the tympanum will be exactly over the line a, b, and c is the center of the cylinder of descent for the weights. The direction of the cord it is supposed may be from d to where an aperture in the wall may pass it on the pulley or point of suspension, thus requiring but a single change of direction. This, however, is for the consideration of the artist.

"The bell is to be suspended on an iron gallows sufficiently strong mounted on the ridge pole of the pediment, perpendicularly over the clock works; no ornament is to be given to it, nothing which may attract notice, or withdraw the attention of the observer from the principal object, somewhat in this simple style. It must be free to be rung independently of the clock.

"The weights of the striking as well as the going parts descend in the same cylinder, but the ringing rope may go down the opposite cylinder at f, which is occupied by winding stairs. The winding up of the clock must be on the back side of the works within the hollow of the roof. And there also means must be furnished of setting the hands."

Th. J.

Monticello, June 4, 1826.

Ten days later Mr. Coolidge sent the following answer:

"DEAR SIR. I have seen Mr. Willard, and given him your order for a clock and bell. In consequence of my conversation with him on the subject sometime since, he procured castings of the principal wheels and made other preparations at his own risk, which involved him in some expense, and make an advance desirable; at present all he asks is 100 dollars, and the work done amply warrants the payment of such a sum. There would be an advantage in furnishing Mr. Willard with money from time to time, as it would prevent his contracting for other work, and enable him to give his whole attention to this, in which case the clock would be finished by September 1st. As for the bell he prefers to select it himself, and thinks one of 400 lbs. large enough. The clock he engages shall be inferior to none in the United States, and he gladly accepts your offer of permitting him to put it up himself, as the accuracy of the movement depends as much upon the skill with which it is put together as upon that with which it is made. Independent of his merit as an artist, Mr. Willard's great respect for yourself makes one very glad that he is to be employed for the University.

"In February last I wrote to thank you, sir, for the desk on which ' The Declaration' was written, but fear that my letter was not received. I mention it lest you should think me either negligent or indifferent — to which charges I plead not guilty. With great respect,"

Joseph Coolidge, Jr.

Boston, June 15, 1826.

The sheet containing the plan and elevation of the Rotunda where the clock was placed, drawn by Jefferson, has survived, and is shown, in Plate 10, largely reduced. Unfortunately, Mr. Jefferson did not live to see the clock, as he died on the 4th of July following. Simon Willard visited Jefferson several times at Monticello, and had many ancedotes to relate of his conversations with him. He was especially delighted to tell how Mr. Jefferson asked him one day to take to pieces a very complicated French clock, and put it in running order. While he was so engaged, Jefferson talked to him about a very important treaty, then pending. Noticing that Mr. Willard did not seem especi-

PLATE 10

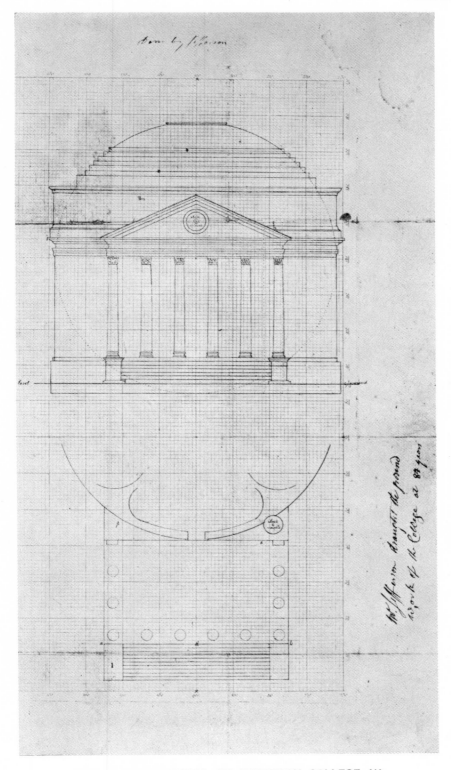

PLAN OF THE ROTUNDA OF JEFFERSON COLLEGE, VA.

DRAWN BY THOMAS JEFFERSON   SIZE OF ORIGINAL 8 x 17 INCHES

ally interested in so grave a topic, remarked, "You do not seem to be impressed with the importance of this matter, Mr. Willard." "Why no," replied Mr. Willard, "I have never studied political affairs and really do not understand them." Jefferson replied with some impatience, "Why, Mr. Willard, every good citizen should be versed in politics and be ready with opinions." "Very likely," returned Mr. Willard; "doubtless every man should be learned and skilful enough to take up any branch of business that is offered to him." Saying which, he rose from the table and prepared to depart. "Don't go, Mr. Willard," said Jefferson, "until you have put the clock together." "Oh," said Willard, "you can do that." "But I cannot," said Jefferson. "Ah," said Willard, "you cannot put the wheels of a clock together, yet you expected me to know all about treaties." The President saw the point, but his answer is not recorded. Later on the President took Mr. Willard out into his plantation, and cut a sapling which he had made into a cane, silver mounted, with an inscription, "Thomas Jefferson to Simon Willard, Monticello," and the date, and presented it to him.[29] Also Mr. Willard was the guest of Ex-President Madison, who treated him with the utmost consideration, and also presented him with a cane, silver mounted, with the inscription, "Presented by James Madison, Ex-President of the United States, to Simon Willard, May 29, 1827." From the date on this cane it is supposed that Simon Willard visited Madison at the time he went to the University of Virginia to put up the clock ordered by Jefferson. Madison was then at his plan-

[29]Lost or Stolen in 1850.

tation at Montpelier, Orange County, Va., and not many miles from Charlottesville, where the University is, and Madison probably gave him the cane at that time. These two canes were Simon Willard's most treasured possessions. He always used one or the other when he went out to

walk. The cane he is holding in his hand in the picture is the Jefferson cane. Simon Willard was quite prominently connected with Harvard College ; he had sole charge of the clocks in the College for many years. His feat in perfecting the great Orrery of Mr. Joseph Pope, who after having devoted several months to the detection of an error in the construction, was compelled to abandon it as a failure, was a fine piece of mechanical skill. This particular Orrery would work all right up to a certain point when suddenly the whole solar system would give a tremendous jump, to the despair of its inventor. Many skilful mechanics were called in to remedy the defect, but all gave it up, and finally Simon Willard was appealed to, with the offer of untold sums if he could make it run smoothly. Simon Willard looked it over carefully, took out his drill, drilled a hole in a certain place, put in a rivet (he always called it a ribbet in telling the story) and the Orrery worked to perfection, the whole operation not taking over an hour. The authorities were delighted. "Now, Mr. Willard," they said, "how much do we owe you." "Oh," said Willard, "about ninepence will do, I guess." He often told of this with great delight.

Simon Willard presented two clocks to the College, one, a Hall clock, stands in the Faculty Room (Plate 11), and has an inscription in Latin on the dial

> S. Willard
> in U sum Coll. Harv.
> Praesidis Successorumque fecit

The other is a large Regulator clock (Plate 11) hangs on the wall in Room 4, University Hall, and also has a Latin inscription.

> Academiæ Harvardianæ
> Ad Bibliothecam Praesidis Ornandum
> Simon Willard
> qui fecit
> Grato animo donavit
> A.D. XIII Kal Sept. MDCCCXXIX

Simon Willard was always very proud of his Latin. In this Regulator clock was found a paper some years ago, a copy of which is given below.[30]

"1829, August 20, Simon Willard of Roxbury, Clock-maker, who for more than fifty years had been employed by the Corporation of the University in the general care and superintendence of the clocks belonging to the institution, this day put up in the library of the President for his use and that of his successors in office, an elegant clock or regulator, of which he asked the acceptance of the Corporation for the use of the President's library, as evidence of the givers grateful sense of the favors conferred on him by the Corporation and Government of the College. It was voted that the thanks of the Corporation be presented to Mr. Willard for this valuable and useful present, and that the President communicate to him their sense of their favor."

[30]By the courtesy of G. W. Cram.   Recorder of Harvard University.

This Regulator clock is a very fine specimen of Simon Willard's workmanship. Many anecdotes are told of his friendship with President Kirkland.

> "He early became a welcome visitor of the Presidents and Professors of Harvard College. Indeed he was the familiar friend of five successive Presidents of the College."[31]

He also had charge of the clocks of the First Church of Roxbury, and made the clocks for the Society.

> "Simon Willard appointed in 1791 to take care of the church clocks and had charge of it for many years."[32] "May 7, 1804, the Parish voted to purchase only one clock for the inside of the Meeting House until the Pews were sold. This clock was made by Simon Willard and is undoubtedly the one still in the church."[33]
>
> "In April, 1806, the new clock with one dial was set up in the tower of the new meeting house by Mr. Simon Willard who made it at a cost of $858."[34] " Mr. Simon Willard continued to have charge of the clocks in 1818."[35]

Simon Willard was also a great friend of Josiah Quincy (1772-1864). In 1826 he made a Timepiece for Quincy who wanted it for a wedding present for his daughter. At the time he made this clock, Simon Willard made a bet with Quincy that he (Willard) would live to be a hundred years old. Quincy took the bet and a document was duly drawn up, signed and sworn to, that Messrs. Willard and Quincy made a solemn compact by which said Willard was to make with his own hands on his one hundreth year a Timepiece of the same kind for said Quincy, the

[31]Edward Holden. *Boston Evening Transcript*. September 4, 1857.
[32]W. E. Thwing. History of the First Church of Roxbury. Page 200.
[33]Ibid. Page 219.
[34]Ibid. Page 220.
[35]Ibid. Page 222.

PLATE 11

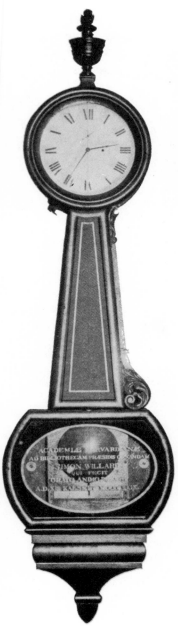

SIMON WILLARD

HALL CLOCK
OWNED BY
HARVARD COLLEGE
CAMBRIDGE, MASS.

REGULATOR CLOCK
PRESENTED TO
HARVARD COLLEGE
CAMBRIDGE, MASS.

HALL CLOCK
OWNED BY
BERNARD JENNEY
SOUTH BOSTON

PLATE 12

SIMON WILLARD

CLOCK IN THE OFFICE OF THE CLERK OF THE SUPREME COURT
WASHINGTON, D. C.

said Quincy to pay whatever price was demanded by said Willard. This compact was frequently spoken of between the two friends, and as Simon approached his ninety-sixth year, it seemed as if he would win his bet, he only missed it by four years. Perhaps at the time of his bet Simon Willard had in his mind his Grandmother, Martha (Clark) Willard, who died at the age of one hundred,[36] and his great Grandfather, Isaac Clark, who died at the age of one hundred and two.[37]

[36] In Memory of the Widow Martha formerly wife of Maj Joseph Willard who died June 3, 1794 in the 100th year of her age.
Having had a posterity of 12 Children. 90 Grandchildren & 226 great Grandchildren & 53 of the 5th Generation.

[37] Here lyes Buried the Body of Capt ISAAC CLARK who departed this life May 26th 1768. Age 102 years.

Here lyes ye Body of Mrs SARAH CLARK Wife to Capt ISAAC CLARK who departed this life May 17th 1761. Aged 88 years.

He lived 70 years with the wife of his youth. His offspring that defended from him was 251.

During the course of his long life Simon Willard had a large correspondence and received letters from distinguished people from all parts of the country. Of these letters he was very proud and at one time lent them to Josiah Quincy to read. The letter following explains itself.

What became of this correspondence which would be simply invaluable now, the author is unable to say. In 1837, Simon Willard was engaged by the United States Government to make two clocks for the Capitol at Washington. These two clocks after being made, were set up and tested at his son's (Simon Willard, Jur.), store at No. 9 Congress

[36]Grave Stone inscription in old cemetery, Grafton, Mass.
[37]From double grave stone in the old cemetery at Framingham Center, Mass.

St., Boston, before being packed for shipment. Simon Willard went especially to Washington to put them up. One (Plate 12) was ordered by Associate Justice Story,[38] and was put up in the United States Senate Chamber, afterwards the Supreme Court. The other clock was a specially construct-

Simon Willard Esq

Dear Sir

I have the pleasure to return you the documents with which you kindly furnished me. They are curious & interesting.

At your advanced period of life, these reminiscences of your early & continued usefulness and of the appreciation of it by your fellow citizens are very justly objects of pride & grateful reflections. wishing that your old age may be as happy as your youth & middle age have been distinguished

I am Very truly your friend

Cambridge 19 August 1840 — Josiah Quincy

ed movement for the case of the clock now in Statuary Hall, the famous allegorical Clock Case, Clio, the muse of History, designed and executed by the sculptor, Carlo Franzoni, in 1819. This clock (Plates 13 and 14) is familiar to

[38]Letter from Elliot Woods, Supt. of U. S. Capitol and Grounds, Washington.

PLATE 13

SIMON WILLARD — FRANZONI CLOCK
STATUARY HALL, U. S. CAPITOL
WASHINGTON, D. C.

PLATE 14

STATUARY HALL, U. S. CAPITOL, WASHINGTON, D. C.

FRANZONI CLOCK

every sightseer who visits the Capitol. The author quotes a letter in regard to these two clocks.[39]

> "In reply to yours of the 20th inst. I have to state that I always had charge of the clocks in the United States Capitol for over forty years, and am thoroughly familiar with the older clocks in use during this time. To the best of my knowledge there are but two Willard clocks now in the Capitol, one in the Chief Clerk's office of Supreme Court and the other in the Franzoni Clock, the latter I have always believed was a specially made movement for this case. You may not recall that the present Supreme Court was up to 1859 in United States Senate Chamber, and the clock now in the Chief Clerk's room is probably one of the two ordered in 1837, the other being the Franzoni clock now in Statuary Hall, and which was at that date, the United States House of Representatives." "P. S. I am aware of the fact that the Franzoni Case was made in 1819, but as affairs moved more slowly in those days than at present it is not unlikely that the case was laid aside and not used until 1837 when the two clocks you refer to were made by Simon Willard."

These clocks Simon Willard put up himself. While he was in Washington he was shown much attention by the President (Van Buren) and the various members of Congress. A letter from Simon Willard to his son describing this trip is given on page 32.

It is curious that the letters about his first and last clocks at Washington, nearly forty years apart should have survived, and is about all the correspondence of Simon Willard, the author has been able to find. This was almost the last important work he was engaged in, and soon after his return to Roxbury he began his preparations to retire from business.

Simon Willard's services were constantly in demand for the manufacture of Church or Turret clocks, and many New

[39]Letter from Mr. Henry C. Karr, Washington, March 30, 1908.

Washington Oc. 29. 1837

My Son

I arrived here friday evening in good health much better then when I left home, yesterday I set up the Clock. I found it complete as it was in your shop. They are much pleased with it. I shall tarry here a few days longer. I have not yet called on the Presidents. the President is not at home he is at the Springs but expected back in a few days. Adams is here shall call on him. —

I spent half day in N york half day in Phil -adelphia half day in Baltimore, hope that I shall return in safty & find all my friends in good health.

with much respect
& love to you all

Simon Willards

I shall stop in N york a day or two, Mr. Leggot gave me a handsome invitation, I did not look up Joseph

Should like to receive a line from you. how your family are Elizza was unwell my Mary respect is pretty well — give my love to every one of my family.

I cannot pay the Postage the Post Office is Some distance you can write me at N York

England Churches had one of them. The author has compiled a partial list of the more important clocks (see pages at end of book) but doubtless many have escaped his notice. The lapse of so many years, repairs, fires, removals, etc., making it very difficult to locate and identify them. About December, 1839, he retired from the business in which he had worked so long and faithfully. From 1840 to 1843, he lived with his son, Simon Willard, Jun., in Boston, then for a couple of years he staid at his son-in-law's (Edward Bird) place, on Boston St., Dorchester, Mass. (now Columbia Road). While staying in Dorchester, he often used to walk to Elnathan Taber's (an old friend and former apprentice) shop on Taber St., Roxbury, and amuse himself by making clocks. Taber's place was about a mile and a half distant.

His remaining years were spent at his daughter's (Mrs. Mary Hobart) place in Milton, Mass., and his son-in-law's (Isaac Cary) residence in Boston. Simon Willard's faculties were retained to the end of his life. His sight and hearing were unimpaired, and at the age of eighty, he read his favorite paper, the *Boston Evening Transcript*, without glasses. He also shaved himself, and without the aid of a looking-glass. He always was very proud of his ability to go about and look after himself.

While visiting the old Benjamin[4] Willard homestead this summer, the author was told the following story by Mr. William Merchant, the present owner of the place. Mr. Merchant said that he bought the place from Mr. Henry Wesson, who said that when he was a young man about

[4] Grafton Vital Records.   Page 148.

26 or 27 years old, a very old man came to the place, and after looking around he approached Mr. Wesson and said to him, " My name is Willard. I used to make clocks here when I was a young man and I wanted to see the place once more before I die, and I have come a long way to see it." Mr. Wesson said the old man looked around for a while and then went away without saying anything more. As Mr. Wesson was born in 1814 this would make the date of the visit about 1843 or 1844. It is possible that Simon Willard was the one who visited the old place. Mr. Wesson died in 1903, aged 89 years, 4 months.

A description of Simon Willard at this time, and his habits was recently given to the author by one of his grandchildren.

"I recall him distinctly, for though he was about 93 years of age, and I was only five when we first met (to my recollection). I see now his little figure sitting in his own arm chair by the window in Aunt Mary's room at Milton. He used to sit in this chair most all day, now and then taking a short pair of steps by which he could reach the clock in the room, and opening it would do some little thing to it probably from habit, rather than from any fixing the clock needed. This clock was one of his own Timepieces. His arm chair we have now, the wood work is so beautifully fitted and the style of the chair so simple and neat that I often think he must have made it himself. My brother tells me that at Milton, Grandfather went to visit Gen. Whitney one day, upon coming home, our man Elijah offered to help Grandfather out of the team. 'Don't help me out, don't help me out,' said Grandfather, 'they will think I am an old man.' He was then in his 94th year. We moved back to Boston, in the fall of 1847, he going with us. He kept his room constantly after this. Father used to mix Grandfather a rum toddy, and put a cracker in it every morning and evening. Grandfather looked forward to these times as the events of the day. The morning he died, Father took him in his accustomed glass. Grandfather could not drink it, and said to Father, 'the old clock has about run down.' These were his last words, he went to sleep, and quietly, and without a sign of distress dozed off into the next life."

He died at the residence of his son-in-law, Isaac Cary, Washington St., Aug. 30, 1848, aged 95 years, 4 months, 27 days.[40]   He was buried in the old Eustis St. Cemetery at Roxbury, and later his remains were removed to Forest Hills Cemetery, where they now lie.

Like nearly all inventors and geniuses, Simon Willard was a very poor business man, and reaped very little benefit from his inventions. He never advertised, and was perfectly content to wait and let business come to him, thinking evidently his reputation was amply sufficient to bring business. He allowed his apprentices, and all the other clock-makers of his time to copy his Patent Timepiece, or anything he invented, and seemingly never thought of prosecuting them for infringing on his Patent. Had he demanded a royalty from them he would have been a rich man. Instead he contented himself by haughtily refusing to speak to, or notice the offender, which being precisely what they wanted, the offenders prospered at his expense. It seems almost pathetic to state that after seventy years of incessant work, he retired from business with the magnificent sum of $500 to his credit, and died a poor man. He made a great mistake in selecting Roxbury as his place of business instead of going into Boston. He probably chose Roxbury because rent was low there, and he always had a perfect horror of high rents. Had he gone to Boston, he would have done much better financially, although perhaps his reputation as a clock-maker would have been no higher. Simon Willard was a most tire-

[40]Grave Stone, Forest Hills Cemetery.

less worker.   He worked twelve and fourteen hours a day. He knew nothing about labor unions and eight hours a day and probably would not have believed it had he been told. He was very jealous of his reputation as a clock-maker, and nothing but the very best of material and workmanship was put into his clocks, and in making a clock he did not consider the money side of the transaction at all, he aimed

to turn out his best work, whether he made a profit or not. It is estimated that during the course of his life he made 1200 eight day clocks, 4000 timepieces, besides machinery for light houses, repairing and improving other clocks,[41] etc.   It may be observed here that Simon Willard, and likewise his brother Aaron seemed to be as determined to stay in one location as his brothers, Benjamin and Ephraim were to roam about.

[41]Edward Holden. *Boston Transcript.* September 4, 1857.

Simon Willard and his brothers seemed to have very little to do with each other. The author has been unable to ascertain that they ever visited each other, or had any business dealings, in their later years at any rate. Z. A. Willard states that he heard Simon Willard mention Benjamin Willard once, the other brothers never. As a clock-maker Simon Willard ranks with the best, and he left as his monument a clock that up to the present day, no person has ever been able to improve.

SIMON WILLARD CLOCKS.

Simon Willard had a genius for clock-making. He was an inventor, a natural born mechanic, and a most wonderful workman. In these days when everything is made by machinery, an account of his methods will perhaps be of interest. Of his early clock-making in Grafton, Mass., very little is known. That he made clocks there on his own account is proved by the existence of little Half or Shelf clocks, called by him, thirty-hour clocks, that are occasionally seen, bearing his name engraved on the dial. Somewhere about 1770, or between that time and the time he took up his residence in Roxbury, he made a number of thirty-hour clocks. They were short, not over twenty-four inches in height, made to stand on a shelf and required to be wound every other day. Externally they were not always attractive, the cases cheaply made of cherry or mahogany. The dials are of brass, whether of domestic manufacture or imported, it is hard to say, but perhaps domestic. The movement was well made and of a design afterwards elaborated into the Patent Time-piece of 1802. The movement plates were very close together allowing a narrow barrel which could take only enough cord to run thirty hours. For some reason the weight was very heavy, much heavier than the subsequent eight day clock weight. Inasmuch as the movement was very much the same as the larger clock, same train, same number of wheels, pinions, same pallets, pendulum, etc., it is difficult to see where the economy of work came in, unless it competed with the tall eight day clock then of universal use, and its

PLATE 15

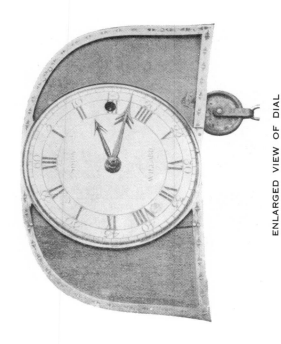

ENLARGED VIEW OF DIAL

SIMON WILLARD—THIRTY-HOUR CLOCK

PLATE 16

ENLARGED VIEW OF DIAL

CASE OPEN

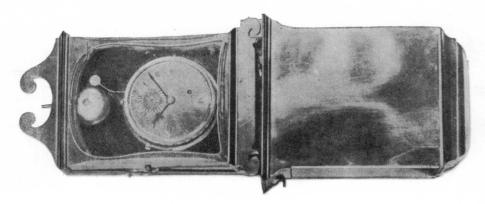

SIMON WILLARD—THIRTY-HOUR CLOCK

cheapness made it desirable. It is doubtful if many were made, and the makers of wooden clocks in Connecticut copied the design and made large numbers of very cheap clocks of this kind. The eight day clock of later design put them out of the market. The annoyance of having to wind a clock daily told against its use and doubtless set the inventive mind of Simon Willard in quest of a design that eventuated in the eight day Timepiece. A few examples of his thirty-hour clocks are shown. One (Plate 15) is about twenty-four inches high, ten inches wide, and four inches deep. The case is mahogany, the dial brass, with the Roman numerals, and has engraved on the dial, "Simon Willard." The dial rests on a cherry wood back surrounded by an engraved brass frame; the weight is of lead, very heavy, and is recessed at one side at the back, to allow the pendulum to swing in; the pendulum rod is of wood.

Plate 16 is a striking clock, about twenty inches high, seven inches wide, and four inches deep. The case is mahogany, the brass dial with a tiny second's hand has "Simon Willard, Grafton" engraved on it in a running hand. The striking hammer is nicely engraved, but the bell is not the original one; the lead weight is heavy, and recessed at the back for the pendulum as in Plate 15. Plate 17 shows by far the most perfect example of this style of clock. It is a miniature Hall clock about two feet high, and is in absolutely perfect condition. The case is mahogany, very finely made. The weight is in front of the pendulum but is not recessed at the back. The dial is of brass and also has the inscription, "Simon Willard, Grafton." There is also a tiny second's hand, the same as the clock in Plate 16. The works are especially good. Taking the clock as a whole, it is a unique specimen of Simon Willard's work at this period, and it possibly might have been made on an order. Still another examined by the author had a Latin inscription, "ab hoc Momento pondet Æternitas" and the name "Simon Willard" on the brass dial. The weight of this clock shows the scarcity of metals in the country at this period, 1770-1780. It consists of rough bars of lead, evidently cast on the premises, mingled with pieces of cast or wrought iron, the whole being tied together like a bundle of sticks, making a very crude-looking weight.

All the movements of these clocks show the nicety of finish peculiar to Simon Willard. He did not make this style of clock after settling in Roxbury, devoting himself to the tall Hall eight day striking clock, Turret or Church clocks, Gallery clocks, and general repair work. The author has never seen one of his Hall clocks marked "Grafton," although Simon Willard is known to have made them there. He made all his clocks by hand, and the methods employed by him were primitive in the extreme. As no fine steel or brass was made in this country, he was forced to import all his material from England. A small forge in which he heated his steel, a hammer and anvil to reduce it to the right size, a file to cut and round up the leaves of the pinion, and hand polishing, were his methods of shaping and finishing the pinions. His brass he was obliged to hammer down to the requisite thickness, and also to give it the necessary toughness, and then finish with the file and polish by hand. The wheels of his clocks he prepared with the utmost care, hammering the brass with slow and even strokes

PLATE 17

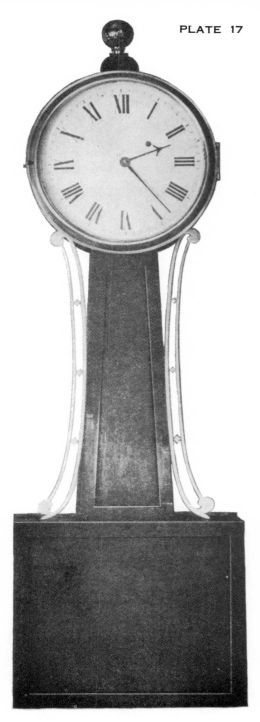

MINIATURE HALL CLOCK
OWNED BY
ARTHUR W. WELLINGTON
BOSTON, MASS.

SIMON WILLARD
TIMEPIECE PRESENTED TO HIS DAUGHTER
OWNED BY
THE MISSES BIRD
DORCHESTER

PLATE 18

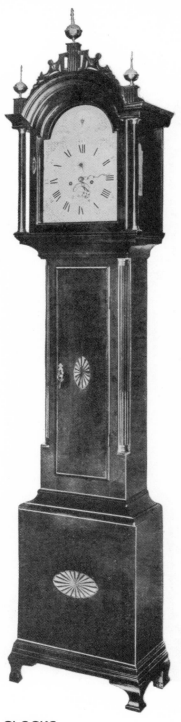

SIMON WILLARD — HALL CLOCKS

OWNED BY
J. T. NEEDHAM
NORTH CAMBRIDGE, MASS.

OWNED BY
HENRY B. MARTIN
MILTON, MASS.

till the utmost degree of toughness was obtained, then with the slitting and rounding up file, he cut out the teeth.

In cutting his wheel teeth, he did not mark out the spaces on the blank wheel and cut the teeth to measure, but he cut, rounded up, and finished the teeth as he went along, using his eye only in spacing, and always

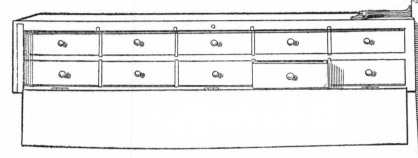

OLD VISE AND TOOL CHEST USED BY SIMON WILLARD

came out even, as accurately, in fact, as the wheel cutting machine, which came into use years later, could do it. It is doubtful if such a feat in mechanics was ever done before, and certainly never since. The 'scape wheel of the dead beat escapement of the clock must be absolutely accurate in the spacing, and it is a very difficult piece of work even with a machine made for the purpose, but all the 'scape wheels and pellets were worked out by hand with no assistance, but the eye and touch. Some of the escapements are as perfect to-day, and do as good service as when they were made a hundred years ago. Simon Willard used to say in regard to his work, that working alone this way, he could turn out complete, an eight day striking clock in six ordinary working days. In his latter days, with more and better tools, he could probably turn out a Timepiece complete in one day, cases excepted of course. Of course in making his large turret clock wheels, he could not hammer out the brass, the blanks had to be cast, but he filed the teeth out from the cast blanks exactly as his smaller clocks. It is not known where Simon Willard got his wheel blanks cast, or by whom, but there is some reason to believe that at one period Paul Revere made some of the castings. As an object lesson in file work it is worth while to examine the works of one of his clocks; one can but admire the beautiful workmanship and finish. In fact, it is a matter of astonishment that he could do such work even when he was over eighty years old. The Turret clock in the Old State House, Boston, made when he was 78, the clock made when he was 82, as a wedding present to his daughter, Sarah Brooks Willard (Plate 19) with the following inscription on the case:

Made
and
Presented
to Mrs.
Sarah B. Bird
by
her father
Simon Willard
in his
82 year

and the clock made for the Capitol at Washington when he was 85, are as beautifully done as any of his early work,

PLATE 19

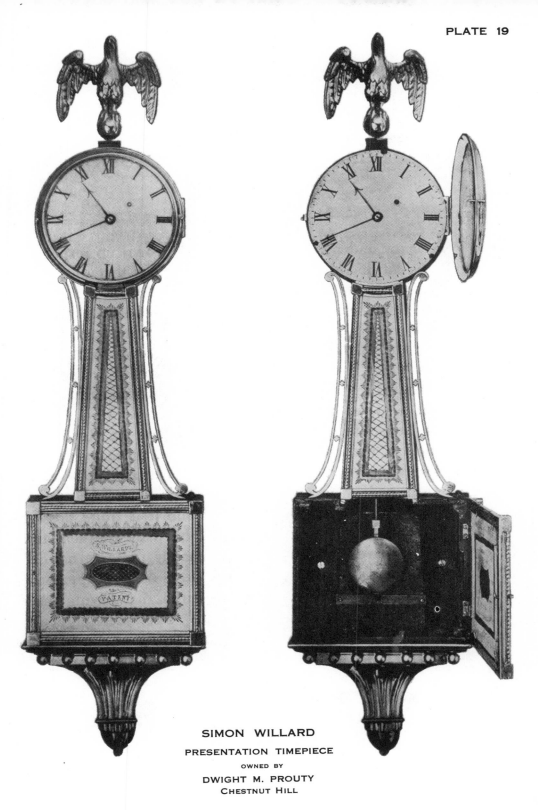

SIMON WILLARD
PRESENTATION TIMEPIECE
OWNED BY
DWIGHT M. PROUTY
CHESTNUT HILL

PLATE 20

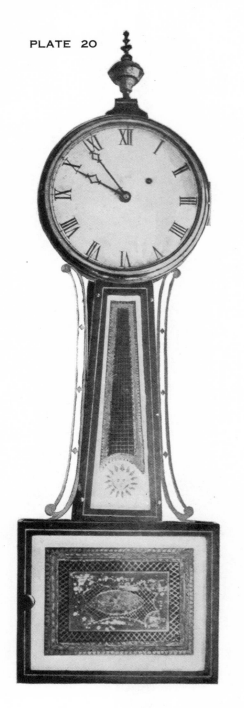

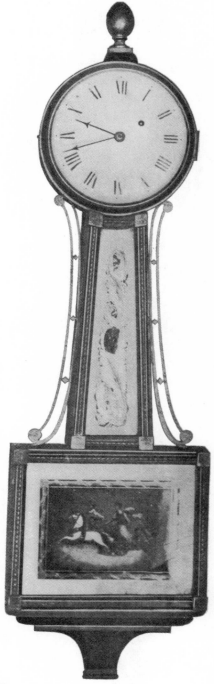

TIMEPIECES

SIMON WILLARD
OWNED BY
R. H. MAYNARD
BOSTON

AARON WILLARD
OWNED BY
Z. A. WILLARD
BROOKLINE

and the fact that these clocks are still running is sufficient proof of his remarkable skill. He made all his clocks this way until very late in life when he occasionally used a machine that had been invented in England for cutting out wheel teeth. He used to say that not one man in ten thousand knew how to handle a file properly. In filing clock wheels, the file would be held so as not to touch the edges, leaving a slight depression in the centre, a square laid across the work should touch both edges, leaving a hollow in the middle. Very few, even experts with the file can do this, they leave an elevation in the centre. In spite of being all file work, the teeth were well made and beautifully proportioned, all the works finely finished and polished. At no time was his assortment of tools very extensive, or for that matter was any of the clock-makers of his time. This can be seen by the inventory of the tools of his apprentice, Elnathan Taber. Taber bought most of Simon Willard's tools when he retired from business. It will be noticed the machine for cutting wheel teeth is specified. The appraisers seem to have put rather a low valuation on the machine. Simon Willard was a most careful and painstaking workman and extremely sensitive about the reputation of his clocks. They were made on honor, none but the best of material was used, and a clock of his making, that has had good care, is as good today as the day it was made. He allowed no piece of work to go out of his workshop without passing through his hands, and undergoing a most exacting inspection. In his tall eight day or Hall clocks, the pendulum rod was made of some selected wood such as maple, oak, or apple, carefully baked in an oven until all moisture was removed, and then given six or eight coats of varnish, each coat being rubbed down. Simon Willard never made his pendulum rods for his Hall clocks of metal, but as these wooden rods were rather apt to get broken, by moving and rough handling it is the exception to find one of his clocks with the original wood pendulum rod. When broken, a rod of metal was always substituted by the repairer. The weights were usually of brass, finely finished and polished and weighted with shot. Frequently weights are found made of tin, painted black, and filled with lead, occasionally of rough cast iron, but these were probably substitutions of a later period.

The dials of both his Hall clocks and Timepieces were painted with from eight to ten coats, each coat rubbed down until the dial face was like polished ivory. This dial painting was a separate trade. Charles Bullard (Boston and Dedham, 1794-1871) was celebrated for this. Simon Willard made the dials of his Hall clocks and Timepieces of heavy sheet iron, and once in a while of wood. He made but very few dials of brass that the author is aware of. Z. A. Willard states that he never saw but one, although it is to be noted he made the dials of his little Half or Shelf clocks of brass. The dials of his Hall clocks were often very finely decorated (Plate 18). Before he employed Charles Bullard, and an unknown English artist, the dial decorating may have been done by John R. Penniman.[1] Simon Willard had the Arabic numerals to indicate the hours in his earlier Hall clocks. The later clocks had the Roman numerals. The cases for the Hall clocks were carefully made of selected well seasoned wood, oak, red and yellow mahogany, or cherry, and were often beautifully inlaid with satin or

[1] R. C. Vol. 34. Page 150.

holly wood. Like the dial painting, clock-case making was a
separate trade. Henry Willard, Roxbury, Mass. (1802-1887),
Charles Crane Crehore, Dorchester (1793-1879), and William
Fisk, Watertown (1770-1844), especially the latter, being notable clock-case makers. The brass ornaments on the hoods of
his Hall clocks were an urn or vase, or bell and spike, rarely
being eagles. Like all the other clock-makers of his time
Simon Willard made the Chime and Musical Clocks. He
always had bells on his striking clocks, *never* gongs. If he
made a Hall clock with a seconds hand, he generally made it
with the dead beat escapement. This particular movement
invented by George Graham, England (1673-1751), is very difficult to make and most clock-makers of that time made the
recoil escapement. Plates 11 and 18 show representative types
of Simon Willard's Hall clocks. Plate 18 is one of his very
early clocks, the author quotes part of a letter from the present
owner about it,[2] "It was one of the first clocks set up by Simon
Willard, he set up the clock himself in Dr. Norwood's house
in Manchester, Mass." The clock is a very handsome one,
rather taller than the average of Hall clocks, case of mahogany,
with brass fluting in the pillars. The dial is a very elaborate
one, handsomely painted, giving the days of the month, and
the changes of the moon. Top ornaments are not the original
ones, and the wooden pendulum rod is lacking. This clock
was probably sold and put up by Simon Willard in one of his
peddling expeditions. Plate 11 is a good average specimen.
On same plate is the clock given to Harvard College by Simon
Willard. Top ornaments are not original. Plate 18 illustrates
the finest specimen of Simon Willard's Hall clock the author

[2]Letter from Mr. J. T. Needham. North Cambridge, Mass.

has ever seen. It is in absolutely perfect condition, with all the original parts, even to the wooden pendulum rod, the case is a very fine one, handsomly inlaid, dial nicely decorated. This clock seems to have had exceptionally good care taken of it, and would certainly date from 1800 and perhaps earlier.

In 1802, Simon Willard brought out his Patent Timepiece, which proved an instant success, aside from its sterling qualities as a timekeeper, its graceful shape made it popular as an ornament, and having painted glass fronts gave the glass painters a chance to show their skill, and some very beautiful work was done by some of these artists. Simon Willard's improvements in his Patent Timepiece consisted, First,—In abolishing the striking movements, thus reducing the number of wheels to the smallest possible number, and thereby making the whole clock movement of the utmost simplicity. Second,—Making the distance between the movement plates wider, thus allowing sufficient cord on the barrel to run eight days. Third,—Placing the pendulum in front of the weight thus giving room for repair and regulation. Fourth,—The heaviness of the weight was somewhat reduced and the weight made longer and narrower. Fifth,—The Pendulum and Guide were placed in front of the movement. Sixth,—The giving an oblong space in the pendulum so that it may swing clear of the centre pinion and hour and minute wheel collars. Seventh,—The calculation of the train in consequence of the shortening of the pendulum. Eighth,—The method of securing the pendulum when transporting the clock. Ninth,— The shape of the case. All these things were original with Simon Willard, and the whole clock was so simplified that after over one hundred years of use, no improvement whatever has been made on the original design, a most remarkable feat in the history of inventions. Neither has any improvement been made in the case, and the whole clock is now made on the same lines patented by Simon Willard in 1802. Simon Willard had a number of characteristic little contrivances on his Timepieces, which are often of assistance in identifying his clocks, but it must always be remembered that everything he did was always promptly imitated by every clock-maker round about. The dial was fastened to the case by three little hooks, slotted so as to be turned by a screw driver, the door was opened and locked by a catch that could be operated by the key that wound the clock, a very neat contrivance, as it prevented children from readily opening the door, and playing with the pendulum. The movement was fastened to the case by two long screws that passed through from front to back, the pendulum bridge was known as Willard's T bridge, and as it was a very nice piece of work, the other clock-makers generally refrained from imitating it. The clock hands were of one pattern of which Plate 23 is a good example. The catch for the bezel case may be mentioned also as another little mechanical contrivance. In addition there is a peculiar nicety of finish to the movements, impossible to describe, that only one thoroughly familiar with Simon Willard's work, can identify. The author is obliged to confess that this last is beyond him, and when in doubt, he is obliged to call on his father, Z. A. Willard, to positively identify a doubtful clock movement. It may be remarked here that Simon Willard could have patented some of the contrivances mentioned above. Of his Patent Timepieces,

Simon Willard made three kinds, one, the simplest, having the entire case made of mahogany, brass bezel case, brass side arms, but without painted front glasses, or a brackett or base piece (Plate 17). The case, however, is sometimes finely inlaid. The second had a mahogany case, brass bezel case, side arms, and with painted glass fronts, and also without a base piece. This style was the kind he most commonly made. The third was of mahogany, enamelled white, with gilded beading, polished brass bezel case and side arms, ornamented base piece, and provided with extra painted glass fronts. The top ornaments of his Timepieces varied somewhat. Simon Willard was very much inclined to a wooden or brass acorn, gilded, or a ball and acorn leaves. (Plate 17). He *never* used the spread eagle.

Simon Willard got hold of an Englishman, an artist, and employed him to paint the glass fronts of his Timepieces, and the decorative work on the dials of his Hall clocks. He had to pay him from $10 to $20 a pair for the glasses, a very high price for those days, but the man was a genuine artist, and his painting was a real work of art. The design was a combination of arabesque and scroll work, with cross hatching, done in gold leaf on a white ground. It resembled the finest lace work, and was exquisitely done, and it is very doubtful if it could be successfully imitated today. The artist, whose name has unfortunately been lost, died or moved away about 1828. Specimens of his best work are very rare. This artist had an apprentice, Charles Bullard, who although he did very fine work, never quite equalled the work of his master. A Simon Willard Timepiece, with painted glass fronts, decorated dial, case and base piece enamelled and gilded, polished brass side arms and bezel case and top ornament, was a very handsome piece of work, and also a very costly thing for those days, and Simon Willard only made them on an order. The author regrets he has been unable to find one of these clocks for the purpose of illustration.

When a daughter of one of Boston's Four Hundred married, it was considered the proper thing to give her one of these clocks (called Gift or Presentation clocks) for her dining room, and they never cost less than $80, but as years rolled on, came other times and other clocks, these beautiful clocks were contemptuously called kitchen clocks, and relegated to the kitchen, where the heat and fumes of cooking soon played havoc with them, being particularly destructive to the enamel of the dials. Probaby the most beautiful Timepiece ever made by Simon Willard was ordered by Josiah Quincy (1772-1864) for a wedding present for his daughter. The case of mahogany was finished off in white enamel with gold beading, basepiece the same. The glasses were wonderfully artistic, painted by the English artist whose name has not come down to us. On the dial inside of the hour numbers was painted a garland of roses and cherubs, and outside of the hour numbers was an inscription " Made for (name of the bride), 1826, by Simon Willard, Roxbury."

Another timepiece, similar to the above, but with a different inscription, was made fifteen or twenty years before, by Simon Willard, for Josiah Quincy for his own use. Careful inquiry among the members of the Quincy family has failed to reveal the whereabouts of these Timepieces, supposing them to be in existence. Plate 19 is the nearest approach to the Presentation or Gift clock that the author has been fortunate enough to find. Its history is, that it was a wedding present to Capt. John White of Randolph, Mass., who married Vesta Dunbar of South Bridgewater, Mass., in 1797, and is now the property of his great grandson Dwight M. Prouty of Chestnut Hill,[3] Mass. The case, side arms, base-piece, bezel case and dial are original, as is also the door glass. The narrow glass front is a reproduction of the original glass which was badly cracked, the gilding has been a trifle overdone in restoring. The spread eagle is not the original ornament, but was carved and put on by his son, Warren White.

The clock movement is wonderfully fine. It will be noticed that the painted glass fronts closely resemble those of Plate 21. Who the artist was who painted these glasses it is impossible to say, except that it is not the English artist's work, but the whole clock is a very fine specimen of Simon Willard's best work. The only things lacking are the English artist's lace-work design glass fronts and the decorated dial. This English artist only made the lace-work pattern for the Gift or Presentation Clocks. He painted other and less elaborate designs for the glass fronts of the cheaper Timepiece. The patterns were simple but very striking, the long glass front generally had a spray of flowers running down the centre of the glass, or oak leaves and acorns, cherry leaves and cherries, or a conventional design (Plate 21) done in gold leaf on a white ground. (See Plates 21 and 22.) The door glasses had either stripes of gold and black on a white ground (Plate 22), gold and green stripes (Plate 21), gold and red or pink stripes (Plate 19), or gold and black stripes on a blue ground (Plate 21). The ground color of this last example is very

[3]From information given by Mr. Dwight M. Prouty, Chestnut Hill, Mass.

PLATE 21

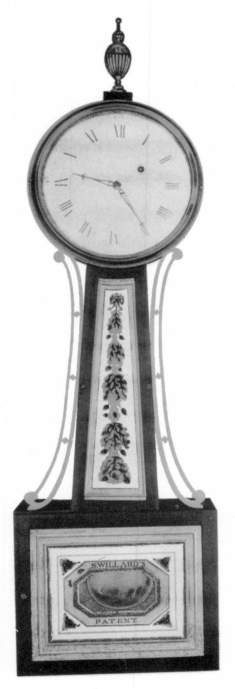

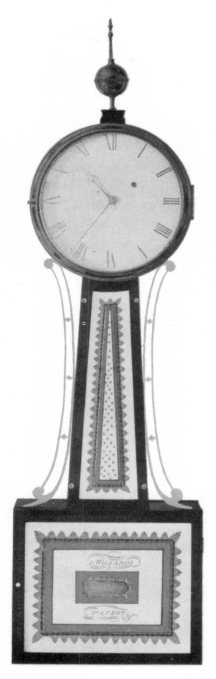

SIMON WILLARD — TIMEPIECES

OWNED BY
THE AUTHOR

OWNED BY
MISS THEODORA WILLARD
CAMBRIDGE, MASS.

PLATE 22

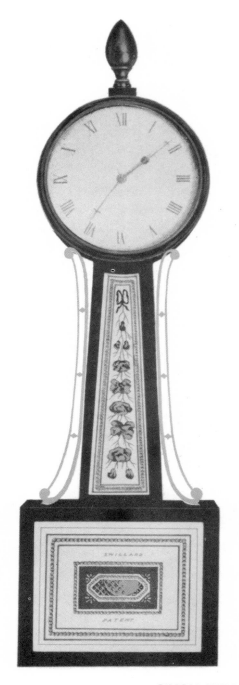
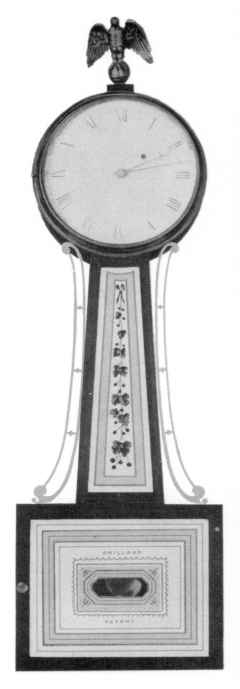

SIMON WILLARD — TIMEPIECES

OWNED BY
FRANCIS H. BIGELOW
CAMBRIDGE, MASS.

OWNED BY
PATRICK MEANIX
ROXBURY, MASS.

exceptional.   The centre of the door glasses had a diamond
or square shaped space to show the pendulum ball.   This was
often finished off with a delicate cross hatching of gold and
black.   The design (Plate 20) is a very unusual one for one of
Simon Willard's glass fronts, and the author is inclined to
assign it to some outside artist.   The work is not suggestive of
either the English artist or Charles Bullard.   The ground
color of the door glass is yellow, the classic head is in gold
leaf.   The author has only seen two examples of this design.
Charles Bullard, the apprentice of the English artist, carried
on the work, and painted the same designs for Simon Willard,
for the author has never seen these particular designs on any
other clock maker's Timepieces, with the possible exception
of Elnathan Taber, as the glasses on his Timepiece (Plate 35)
seem to be Bullard's work.

   With the exception of his Gift or Presentation clocks,
Simon Willard never used gilding on his Timepieces.   Simon
Willard made a large Timepiece called a Regulator clock, for
Banks, Offices and Observatories.   Most of the old Boston
Banks had had one of these Regulator clocks, a few of which
survive.   They were very finely finished, and had polished
brass weights, and as a rule, solid mahogany cases.   A very
fine specimen of his Regulator clock is in the basement of the
Provident Savings Bank, Temple Place, Boston, Mass., another in Room 4, University Hall, Harvard University (Plate
11).   He also made a Gallery clock for Churches and Public
Buildings, and Turret clocks for Church steeples.   His Gallery clocks were fine pieces of work, the whole case was full
gilded, the pendulum case had a finely painted glass front,
and on the top of the clock was a large carved gilded eagle

with outspread wings holding in its beak two strings or fes-
toons of gilded balls or beads.   The gallery clock in the First
Church in Roxbury, Mass., is a beautiful specimen of this
particular style of clock (Plate 23), and the door glass is
undoubtedly the English artist's work. Another Gallery
clock almost identical is in the Second Church of Dorchester,
Mass.   These large Regulator clocks were made with the
dead beat escapement, and very often the Timepieces.

Simon Willard always had good taste in the make up of
his clocks, particularly in his Timepieces. In fact, their
simple elegance always strikes the critical observer. Like all
other clock-makers, Simon Willard occasionally made a freak
clock. He attempted a forty day clock. It had a mahogany
case about eighteen inches high, with brass feet, on the Half
clock model, fitted with a small brass dial, and a very heavy
weight, the whole idea was bad, as the weight being so heavy
and having such a short distance to fall, it called for a great
many teeth on the wheels, and a great many leaves on the
pinions, which could not stand the strain and the whole
mechanism wore out quickly. He made another on the same
plan with striking attachment, surmounted with a bell glass
with a knob to lift off. It was known as the Eddystone Light
House clock from its shape (Plate 24), and was not a success.
Very few of either of these clocks were ever made. These
clocks are mentioned simply to show how inventors incline
to complicated machinery.

It was the custom of the clock-makers of old times to spend
their winters making clocks, and as soon as the roads were
good in the spring, would load these clocks on a wagon,
and ride around the country peddling their clocks, and at the

PLATE 23

SIMON WILLARD
GALLERY CLOCK
FIRST CHURCH OF ROXBURY
ROXBURY, MASS.

PLATE 24

SIMON WILLARD
EDDYSTONE LIGHTHOUSE CLOCK

same time turning an honest penny repairing and regulating such clocks as they found out of order along their route. Simon Willard had his beat along the North Shore, Aaron Willard along the South Shore. It would be interesting to know how they managed to take a load of clocks without damage over such poor roads as were then in existence. The author has traced Simon Willard in his peddling expeditions as far as Bangor,' Me. He seems to have stopped this way of selling clocks about 1805.

'Statement of Z. A. Willard who heard Simon Willard once say he went to Bangor, selling clocks.

CORRECTION OF ERRORS
REGARDING THE LIFE AND WORK OF
SIMON WILLARD.

So many and persistent errors have grown up respecting Simon Willard, his brothers, and his clocks, that the author has thought it advisable to devote a chapter for the purpose of correcting them. It has been asserted so often that it has received general credence from all who have interested themselves in the matter that Simon Willard had as partners, his brothers Aaron and Benjamin. He had apprentices, many of them, whom he instructed in his own careful painstaking way, but he never had a partner. He could not have had his brother Benjamin, for when Benjamin came to Roxbury in 1771, Simon, who was still serving his apprenticeship, was only seventeen years old, and would not attain his freedom till he was twenty-one, which would be in 1774. An apprentice could, however, buy his freedom and it is quite possible Simon did so, for we find clocks with his name on them, thus "Simon Willard, Grafton." Unfortunately there is no date to any of them. There is a family story derived apparently from Simon Willard, after he had retired, that having made with his own hands, without assistance, a complete eight day striking clock, he grew impatient of serving and purchased his time.

In 1776, we find Simon Willard, still in Grafton, married, and in business, while Benjamin seems to have left Roxbury during the early years of the Revolution, as we find no mention of him there after 1774. Moreover, Benjamin, in his various advertisements in the newspapers makes no mention of a partner. That Simon Willard was in business with his brother Benjamin may be dismissed as utterly improbable. Simon Willard came to Roxbury in 1780, and lived in the same house there, until his retirement in 1839. There was his shop, all his business was transacted there, and his family born and raised, and in his very rare clock advertisement (Plate 4) he always speaks of himself, and no partner is ever mentioned, all of which effectually disposes of the story that he was in partnership with his brothers Benjamin and Aaron. It may be mentioned here that Aaron Willard never speaks of a partner.

Another error that has crept in of late years, is that the Willard brothers came from England. It has been stated as fact by several writers, and many times in the newspapers, that Benjamin, Simon, and Aaron were born in England, served their apprenticeship in clock-making there, and came to the Colonies, settling in Roxbury, Mass., when they were young men. It is very puzzling to know how such an error originated, especially when the Grafton, Mass., Records are so accessible, unless people confused Grafton, Mass., with Grafton, England, there being several places of that name there, and mistook Simon[1] Willard, the founder of the family in America, for Simon[5] Willard. This last is likely to be the case, as the author was at one time shown a clock which the owner solemnly assured him, was brought over in 1634 by Simon[1] Willard, the clock-maker. The genealogical notes and documentary evidence given in other chapters are sufficient to refute the absurdity of this story.

It has also been asserted that Simon Willard made

wooden clocks. *Never* at any period of his life did Simon
Willard, or any of his brothers, make other than brass clocks.
The Willards of Ashby, Mass., made wooden clocks, and
doubtless people have confused them with Simon Willard.
It is also claimed that Simon Willard was not the inventor
of the Timepiece. The best possible proof is the patent
issued to him in 1802, and nothing further need be said about
it. It speaks for itself. Still another mistake is made in
calling this Patent Timepiece, a Banjo Clock. This is an
error of recent years. It is called a time-piece in the patent,
advertised as a timepiece, and sold as such, and how or when
the name originated, it is hard to say. Britten in his book
makes this mistake.[1]

Then a vast amount of confusion exists as to the name
on the dials of Simon Willard's clocks. On the dials of his
Hall clocks he put, in a running hand, his name, "Simon
Willard, Roxbury," or "S. Willard, Roxbury," *never* Boston.
Sometimes he had his name engraved on the seconds hand
of his Hall clocks, instead of on the dials, but it was rarely
done. The author has only found two instances of it. Occasionally among his early Hall clocks is found the name
"Simon Willard" in capital letters instead of in a running
hand, and in one instance, the name was in old English
lettering (Plate 18). He never had his name on the dial of
his Timepieces, that the author is aware of, except the Gift
clocks, which would read: "Made by Simon Willard, Roxbury, for ...." On the dials of his large Regulator clocks
he sometimes put his name "Simon Willard." Occasionally in
his later years, when engaged on a large clock he was some-

[1]F. J. Britten.   Old Clocks and Watches and their Makers.   2nd. Edition.   Page 718.

times assisted by his sons, then he would put the name "Simon Willard and Son" on the dial, but it was very rarely done. Two only marked thus have come under the author's notice. One is the large Regulator clock in Room 4, University Hall, Harvard University, the other is the "Franzoni Clock" in Statuary Hall, U. S. Capitol, Washington, D. C., and these might mean either of his sons, Simon, Jr., or Benjamin F. Willard. On the glass door fronts of his Timepieces he had the words: "S. Willard's Patent," in a running hand or capitals, beautifully done in gold leaf. Occasionally he had the words: "Willard's Patent" without the initial "S," on the door glass. Such instances are rare, and seem to belong to his earlier Timepieces, perhaps before he employed the Englishman and Charles Bullard. This was imitated by other clock-makers, but for some reason, probably because they did not dare to imitate too closely, rarely put it on the door glass, rather putting it on the narrow glass front, and this work was very poorly done, the lettering being painted in most cases, instead of being in gold leaf, and having only the words: "Willard's Patent."

A word of explanation is necessary here in regard to the designs on the glass fronts of Simon Willard's Timepieces. He never used the design of a naval battle, American flag and eagle, or a landscape. These designs were peculiar to the Aaron Willards and other clock-makers. Simon Willard employed two persons only, that, the author is certain of, the Englishman and Charles Bullard, and the designs were such as are shown in Plates 19, 20, 21, and 22. In the course of time owing to accidents, etc., many of the glass fronts of his clocks got broken and the owners had to put in new glasses, which were the designs used by the Aaron Willards, etc., as no artist seemed to be able to imitate the work of the Englishman and Charles Bullard. Of course this substitute work takes away largely from the value of a Simon Willard Timepiece. The use of brass spread eagles as ornaments on Simon Willard's Timepieces and Hall clocks is a certain indication of their not being the original ornaments. Simon Willard used the acorn on his Timepieces. The brass spread eagle is of much later date, being imported by a firm in Dock Square, Boston, and sold in great quantities, so when the original ornament on a Willard Timepiece was lost, the brass eagle was almost universally used to replace it.

After Simon Willard retired from business in 1839, Elnathan Taber, his best apprentice, bought most of his tools, and the good-will of the business; continuing the clock-making. He received permission to put the name "Simon Willard" on the dials of the clocks he made. As he made a most excellent clock, Simon Willard, Jr., took all he could make, and sold them at his store at No. 9 Congress Street, and all clocks sold from there had the name "Simon Willard, *Boston*" or "Simon Willard and Son, Boston," on the dial. A Timepiece Taber made had glass fronts painted to imitate mahogany. These were sold for $16.00 and were so popular that Simon Willard, Jr., took all Taber could make. All this has resulted in a great deal of confusion, many people thinking they had a Simon Willard, Sr., Timepiece, when it was really one of Taber's. It is very doubtful whether Simon Willard ever made the striking Timepiece. There is no record that he ever did so. Z. A. Willard states that he never saw or heard of one, although it is barely possible he might

have made them on an order. He never did so on his own
account. The principle was against the simplicity of his
patent and striking Timepieces are continually getting out
of order. Furthermore, the name "Willard" on a clock is
no proof that Simon Willard made it.

As early as 1780 his reputation was so high that other
clockmakers put the name "Willard" on their clock dials,
very much as the Dutch clock-makers put the names of celebrated English clock-makers, such as Tompion, Quare, Lamb,
etc., on their clock dials.[2] The clock thus marked might be,
and very often was, a good one, but Simon Willard never
made it. The author has found several instances of this kind.
Simon Willard clocks, both Hall and Timepieces, have been
counterfeited innumerable times, and it requires a thorough
knowledge of the peculiarities of his workmanship to detect
the fraud. Counterfeits are often made by taking an old
Howard clock movement, having a case made for it, and by
judicious smoking, given the appearance of age; then having
glass fronts painted, the whole is put together, and sold as
a genuine Simon Willard clock.

Just here the author desires to call attention to the very
mistaken method of restoring Willard Timepieces. The
practice nowadays in repairing them, is to cover the entire
front of the case with a heavy coat of gilding, add an elabo-
rately ornamented base piece or bracket, also heavily gilded,
insert new front glasses, even if the original ones are whole.
A spread eagle is put on for a top ornament. Thus restored the
clock is a blinding mass of gilding, very gorgeous certainly,
but overdone. This is all wrong, and the fashion is one to

[2]Britten. Old Clocks and Watches and their Makers. 2nd. Edition. Page 546.

be deplored.   The Timepiece was never that way originally.
Simon Willard, and the Aaron Willards, never used gilding
in such reckless profusion, and would be vastly astonished if
they could see their "restored" clocks.   Such elaborate gilding never entered into their calculations, there was no demand
for it, and it would have made their Timepieces much too
expensive for their customers of those days, who wanted
moderate priced clocks.   The author's father, Z. A. Willard,
states positively that during the entire time he was in the
clock business from 1840 to 1870, during which period hundreds of clocks were brought in for repairs to his store at No.
9 Congress Street, Boston, he never saw a Willard Timepiece,
or any of the old clock-makers' Timepieces, gilded, as the
restored Timepieces are today.   When and how the fashion
for gilding originated, it is hard to say, except that it is recent.
Why people should want Timepieces "restored" this way, is
a puzzle.   If the "restorer" would confine his efforts to keep
the Timepiece as near as possible to the original, the result
would be more satisfactory, and in better taste.   Simon Willard in his Gift Clocks used gilding in moderation, and the
Aaron Willards had only a narrow beading gilded, on their
best clocks (Plate 20).

A little may be said here about clock glasses, and the
artists that painted them.   The author is not certain whether
Simon Willard originated the idea of painted glass fronts for
clocks, or if he got his idea from seeing the paintings on
looking glasses.   At all events, the custom does not seem to
have become prevalent until some time after the Revolution.
After that period, a number of artists, miniature picture, and
ornamental painters came to the United States, a number

settling in and around Boston. It is not easy to identify them in the old Boston Directories, the description being very indefinite, merely giving the occupation as painter, leaving the reader in doubt whether the person was an artist or only a house painter.

Drake[3] says that a John *Ritts* Penniman was at one time employed by Simon Willard to paint for him. Information about this Penniman is very limited. Drake[4] says he lived in Roxbury, and had his shop where Webster Hall now stands. When he was born and where he came from, the author is unable to ascertain. There is nothing about him in the Roxbury or Boston Records. He moved to Boston in 1806,[5] where he lived principally at 57 Warren Street, giving his name as John R. Penniman, painter, or ornamental painter, until 1828 when his name ceases to appear, leaving one in doubt whether he died or moved away. He appears to have been a friend or associate of Gilbert Stuart, the artist, for in the records of the Hollis St. Church[6] is found an entry giving the death of "Gilbert Stuart Penniman, son of John Ritto Penniman, Oct. 11, 1812," no wife mentioned. Whether Drake or the Hollis Street Church Record gives Penniman's middle name correctly the author is unable to say, and is also unable to identify any of the work Penniman is said to have done for Simon Willard.

Mention has also been made in previous pages of an English artist that painted clock glasses for Simon Willard.

[3]R. C. Vol. 34. Page 150.
[4]Ibid.
[5]Boston Directory for 1806.
[6]Hollis St. Church Records. Deaths. 1812. Page 285. Manuscript Copy. City Clerk's Office, Boston.

The most rigid search has failed to ascertain his name, although there is a possibility that it was the Robert Fenwick whose name appears as witness on the Simon Willard's patent (Plate 9). The author has been unable to find any mention of this Robert Fenwick in the Roxbury or Boston or Dorchester Records. The most that can be said is that he came to Roxbury some time after the Revolution, and became acquainted with Simon Willard, who, knowing a good thing when he saw it, employed him to paint clock glasses for him, and it is possible that the Englishman might have suggested to Simon Willard the idea of decorating clocks with painted glasses. Certainly the man was an artist of merit, and his work was very beautiful. Plate 22 (left) is doubtless an example of his work. Unfortunately the photograph does not show the beauty of the work. His finest work was done on the Gift or Presentation Clocks.

The author's father, Z. A. Willard, states that one day a clock came into the store at No. 9 Congress St. for repairs, and he was much impressed with the extreme beauty of the painted glasses, and asked his grandfather, Simon Willard, who painted them. "Oh," said Willard, "those are the old Englishman's glasses." Being only a boy at the time he never thought to ask his grandfather the name of the artist, and has frequently lamented that he failed to do so. The finest work the artist ever did was on the Quincy Clocks in 1826, before alluded to (see Page 49). This was the last work done, Z. A. Willard states, that he is certain of, and thinks the artist died not long after, about 1828. Can it be possible that John R. Penniman and the English artist were one and the same person? This artist doubtless had many apprentices, but only one is absolutely identified, Charles Bullard, who never quite approached the perfection of his master's work, although in the painting of clock dials he was a past master. When so many clock-makers were in business in Roxbury and Boston between 1800 and 1844, it is not possible that all the glass painting was done by Charles Bullard and the English artist. There were many others, and some of them were good ones, but the author cannot identify any of them with certainty.

John[6] Mears Willard, the eighth son of Simon[5] Willard and his second wife, Mary (Bird) Willard, was an artist and excelled as a copyist. His skill in this respect was extraordinary. He might possibly have painted glasses for his father, but he died so young he did not accomplish enough to give him a name or reputation. The author has never seen any specimens of his work. Lewis Lauriat is said to have done something in this line. At first, the artists used a landscape design (Plate 32), a classical figure (Plate 20), American Flag, etc. The war of 1812 was a great windfall for the artists, and the fight between the Constitution and the Guerriere was painted in endless variety, and some were very fine specimens of this class of work. Plate 35 is one of the best. The naval battles of Lake Erie and Lake Champlain were sometimes painted but owing to the large number of vessels required they were not so popular with the artists. The Lake Champlain painting is much sought after.

As before stated Simon Willard never used landscapes or figures on his clock glasses, but adhered to the designs illustrated in the preceding pages. He must have had some special arrangements with Bullard and the English artist for the exclusive use of these designs, for the author has never found them on any other make of Timepiece. About 1850, the clock glass painting seemed to decline in quality for some reason, and between that date and 1872, the art was at a very low ebb, the work turned out being atrocious. Z. A. Willard states that the work was so bad that the Timepieces made for his store had to have the glass fronts painted in imitation of mahogany, owing to the lack of good artists.

PLATE 25

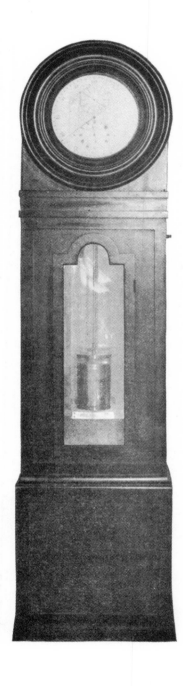

SIMON WILLARD, JUNIOR   BENJAMIN F. WILLARD

ASTRONOMICAL CLOCKS

OWNED BY   OWNED BY

HARVARD OBSERVATORY   F. G. MACOMBER

CAMBRIDGE, MASS.   BOSTON, MASS

PLATE 26

FROM A PAINTING ON IVORY. IN POSSESSION OF FAMILY

J. Willard

Born
January 9th 1795
Taken
December 3 1870

SIMON WILLARD, JR.
WEST POINT. CLASS OF 1815. NO. 125.

Simon [6] Willard, Jr., the subject of this sketch, the second son of Simon [5] and his second wife, Mary (Bird) Willard, was born in Roxbury, January 13, 1795.[1] He inherited his father's mechanical faculty to a remarkable degree. A very notable instance of his ability as a mechanician was exhibited in his extraordinary feat of mastering in eighteen months of New York apprenticeship, so complex and delicate a business as the manufacture and repairing of marine chro-

nometers and watches. While he never made clocks, it was because factories were making clocks in such quantities, and competition had become so keen that there was no money in the business. Had he done so, however, he undoubtedly would have had as great a reputation as his father. His boyhood was a hard one. In his late years, he was very much disinclined to speak of his early days. He had to work hard, early and late, and was insufficiently clothed. He particularly described how he used to suffer from the cold during the winters. He said once to

[1]Roxbury Births. 1632 to 1844.

a friend who spoke to him about the good old times, "I do not want to hear or speak about the good old times. I lived in them, and know they were nothing but hardship and suffering." No better idea of his life can be given perhaps than in his own words, taken from an article written by himself in 1872. He says,

"Till the age of ten years, I had no regular education, but such rudiments as could be picked up at home, or at primary schools with the aid of the New England Primer. At the age of ten I went to the Public Grammar School of Roxbury, kept by Dr. Prentiss, stayed there four years. The youths of that day were not troubled with over education. Such knowledge as was given them was riveted to their memories by the rod and ferule. Moral suasion was not part of the educator's plan. The only School books used were the Columbian Orator, American Preceptor, Murray's Grammar, not the unabridged, but a matter of ten pages, Pike's Arithmetic and Morse's Geography. Left school at the age of 14 years and entered my father's shop, and worked at clock-making for a time till I was apprenticed to a watch maker by the name of Pond in Portsmouth, N. H., with whom I stayed till the war of 1812 was declared. Pond failed in 1812 and I came home and stayed with my father till 1813, when I met a fellow townsman named George Blaney who asked me how I should like to go to West Point. The idea meeting favor we both got cadet warrants through General Heath of Roxbury, a General who served in the Revolutionary Army, and had some influence in Washington."

His life and experiences at West Point will doubtless seem very primitive to the modern West Point Cadet, and his description may prove interesting.

"When I entered in 1813, was eighteen. No examination necessary, received warrant from Secretary of War and was ordered to report at West Point. No oath of Allegiance required. It would be difficult to recognize the West Point of 1813 in the West Point of 1872. The families of such of the Professors as were married resided in West Point in private houses. The town of West Point had no existence. There was a small country store at Buttermilk Falls where Cadets sometimes made purchases. Cadets were allowed two rations of twenty cents each per day and $28.00 per month, and had to find themselves in food and clothing. Four of us attempted to keep house in order to economize, the attempt failed. Reported at West Point about September 1, 1813. Stay at West Point was not continuous as the Cadets at that time were dismissed for the Winter there. Passed the Winter of 1813 in New York City. Passed Winter of 1814 in West Point in company with one cadet named James R. Stubbs, suffered very much from cold, had to lie on the floor wrapped in blankets, and our feet to the fire. Our fires were made from rail fences and such dry stuff as we could find in the neighboring woods. We had to cut and haul it ourselves and sometimes pack it on our backs. We had to forage for a living. It was very difficult to obtain food, we had to buy it of the neighboring farmers and cook it ourselves. It was a very hard Winter.

"In 1814 being desirous of making some fireworks for the 4th of July, I applied to Captain Partridge for permission to make some. After informing him what I knew about fireworks he acceded to the request and I went to work assisted by Cadets Eveleth, Talcott, and Partridge, nephew of the Superintendent. We took possession of the Laboratory, a small wooden building standing where the Mess House does now, and containing three kegs of powder and a great number of cartridges for muskets and cannon. We prepared rocket composition which Talcott was ramming into the cases, when to our inconceivable horror, it took fire. The celerity with which we abandoned that position would have won us great fame in a larger field, and the small space occupied by our bodies during our flight would have made the fortune of an acrobat. Finding after limited flight, that West Point and all thereunto appertaining was not blown into the Hudson, we stopped, turned back and extinguished the fire with a few buckets of water, but the risk was very great. Young Partridge made a detour of three or four miles and brought up in his Uncle's cellar where he was found some hours afterwards nearly dead with terror. That particular 4th of July was never celebrated.

"Graduated March 2nd, 1815, left September, 1815. Commissioned in 1815 in the ordnance Corps and ordered to the Pittsburg Arsenal on the Alleghany River, Pa. Post in command of Major Abram R. Wooley. My first introduction to the service was seeing Major Wooley and Captain Wade eject a man from the U. S. Premises. The individual not having an overwhelming respect of military authority of the United States, drew a pistol and blazed away at the Major, who in return, drew his sword and made a frantic lunge at the Philistine, but unfortunately with more wrath than judgment, so that his weapon passed through Captain Wade's hand. The rage of the combatants being abated by the accident, the individual was allowed to retire with all the honors of War. The duties of the post consisted in inspecting and proving cannon and inspecting shot and shell, all of which were made at the Pittsburg Foundry, also inspecting powder, much of which was damaged, but dried and re-vamped so as to be nearly as good as new. We distributed these arms and the ammunition through the West to all Government forts and our posts. At this time an Armory was building and a repair shop for small arms.

"Arrived in Pittsburg in September, 1815, and sent in my resignation in May, 1816. Resignation accepted and I left the post the same month. Major Wooley was very anxious I should remain in the service, but I did not find it congenial to my taste. The war was ended, there was no active employment of officers, the post was very dull, the daily routine irksome, promotion would necessarily be very slow, and I thought I could satisfy myself better in some other occupation. Returned to Roxbury at age of 21. In 1817 went into the crockery ware business in Roxbury and continued in it until 1824 when I failed. Settled with my creditors paying 40 cents on the dollar, and my notes for the balance, with interest added. These notes I paid out of my subsequent earnings as watchmaker."

"December 6th, 1821, I married Eliza Adams, eldest daughter of Zabdiel Adams of Roxbury. Seven children were the result of this marriage, two now survive."

Simon Willard, Jr., married December 6th, 1821,[2] Eliza

[2]Roxbury Marriages. 1632 to 1860.

Adams. She was born in Roxbury, July 16th, 1795,[3] died in Boston, January 29th, 1881.[4]

Children of Simon and Eliza (Adams) Willard.

Infant son born and died, Roxbury, Sept. 13, 1822.[5]
Mary born, Roxbury, Sept. 27, 1823.[3]
Mary died, Boston, April 17, 1845.[5]
Zabdiel Adams born, Roxbury, Jan 22, 1826.[3] Living.
Simon died, Roxbury, Sept. 1, 1829.[5] Ae 11 mos.
Eliza Josephine born, Roxbury, May 14, 1831.[3]
Eliza Josephine died, Boston, July 13, 1851.[5]
Simon died, Jan. 21, 1837.[5] Ae 3 days.
Helen born, Boston, Oct. 21, 1838.[6] Living.

"After my failure in 1824, went into my Father's clock-making establishment and remained until 1826. In July 1826, at the age of 31, I went to New York and apprenticed myself to Mr. D. Eggert, a very ingenious mechanician to learn the Chronometer and Watch business. Mastered the business and returned in 1828 and set up for myself at No. 9 Congress Street, Boston, where I remained till January 1, 1870, a period of 42 years. I was 31 years of age, had a wife and two small children when I determined to go to New York, learn a new trade and begin life again. All my friends tried to dissuade me. Everybody threw cold water, numberless wet blankets were offered, dismal predictions of failure were made, but I had made up my mind to go and I went.

"On my return, I hired on my own responsibility at a rent of $350, the store No. 9 Congress Street. My Father on hearing it, denounced the transaction as perfectly reckless and one that would precipitate ruin upon the projector. My friends rallied as one man with the chorus of: 'You can't succeed.' I had six and a quarter cents in my pocket, large debts hanging over me, but I went to work, paid my debts and succeeded."

[3]Roxbury Births. 1632 to 1844.
[4]Boston Deaths for 1881. No. 769.
[5]G. S. Mt. Auburn Cemetery.
[6]Family Record.

Simon Willard, Jr., evidently did not have much faith in advertising. The following is the only advertisement the author has been able to find in the papers of that period and he evidently considered four insertions ample.[7]

When he first started in this business he lived at his father's house in Roxbury, and used to walk in every morning to the store, and have it all opened up and ready for business by seven o'clock, walking back again at night. He did this until 1839 when he took up his residence at the corner of Summer and South Sts., afterwards, in 1845, living at 16 Kingston St., Boston.

Probably every old merchant of Boston, and every sea-captain that sailed in and out of Boston Harbor, now living

[7] *Columbian Centinel.* Boston. Saturday Morning, March, 1828.

will remember Simon Willard's store at No. 9 Congress Street. Established during the heyday of American shipping, it continued through and survived its fall. A glance over his books shows names of nearly every shipping merchant in Boston, with many from other ports, who at one time or another bought their chronometers there, or had them regulated or repaired at Willard's store. Likewise appear the names of nearly all the famous ships and their equally famous captains that sailed them. On the day of sailing all these captains called at Willard's store to get their chronometers, which had been rated, that is their daily error ascertained, have a friendly chat, and find out what the weather was to be, for Mr. Willard was famous among sailors as a weather prophet.

Shortly after Simon Willard's return from New York in 1828, he made an astronomical clock of such excellence and rare accuracy that for forty years it was the standard of time for all New England. This clock (Plate 25), known as the Simon Willard Regulator or Astronomical Clock, was the result of the labor of Simon Willard, Jr., about 1832. It has the Graham dead beat escapement, jewelled in rubies. None of the holes were jewelled, but the quality of the brass and steel was such that after forty years of service there was no perceptible wear to any part. The pendulum rod was steel, sustaining a glass cylinder holding sixty pounds of mercury. Its cost was $1000. The works were enclosed in a very fine mahogany case made by Charles C. Crehore. It was presented to Harvard University for the Observatory in 1894 by his son Z. A. Willard and is still in daily use at the Observatory. In a letter to the author from Professor

Edward C. Pickering, May 14, 1909, he says, "The clock has been steadily in use, and has been of great assistance to us."

For many years he had entire charge of all the public clocks of the city of Boston. His Astronomical Regulator, tested by daily transit observation, was the standard time for all the railroads in New England. In 1850, his son, Z. A. Willard, was admitted to partnership.

"He was for many years President of the Cary Improvement Company and the Boston Chess Club. He succeeded General Thayer as President of the Association of Graduates of West Point Military Academy. At the time of his death, he was the oldest graduate of that institution. He presided at the dinner given by the Alumni at West Point in 1873, when President Grant and General Sherman were present and again in June, 1874, which was his last appearance in public. After his return from West Point in June his health failed rapidly and he finally succumbed to the asthma, a

Advertisement used by Simon Willard, Jr.,
to put inside watch cases.

disease which had affected him for over forty years. He died at his residence, 17 Beacon Street, August 24th, 1874, in the eightieth year of his age, honored and respected by all that knew him."[8]

[8]From a memorial written by Major General George W. Collom, also see *Boston Advertiser* of August 26, 1874.

ZABDIEL ADAMS WILLARD.

Zabdiel[7] Adams Willard, only son of Simon[6] and Eliza (Adams) Willard, born in Roxbury, Mass., Jan. 22, 1826. Entered his father's store as apprentice in 1841. Admitted as partner in 1850, and quickly took the lead in the clock, chronometer, and watch business. Suggested and had manufactured in London, England, the celebrated Frodsham Watch, the most remarkable time-keeper ever made by hands, and made his name throughout the United States as expert and authority in that line of

business. In 1855-56 he delivered a series of lectures on Horology and ancient methods of computing time. Retired from the watch and chronometer business in 1870. Married Nov. 6, 1851, Lucy Allen Ware, eldest daughter of the late Dr. John Ware of Boston. The inventive faculty seems to have descended to the third generation, for Z. A. Willard invented many processes, furnaces, and machines for the reduction of ores of gold and silver, worked mines in Colorado and California, acted as physician in mining camps, as well as chemist and assayer.

BENJAMIN FRANKLIN WILLARD.

Benjamin[6] Franklin Willard, the fifth son of Simon[5] and his second wife, Mary (Bird) Willard, was born in Roxbury, November 2nd, 1803.[1] Like his father he was a natural born mechanic and inventor, and a superb workman. His early life was very much like his brother Simon's. He received a very limited schooling, and at an early age entered his father's shop where he learned the clock-making trade. He did not engage in clock-making himself, but worked for outside parties and at times for his father. He was for some time secretary of the Boylston and India Insurance Companies[2] and in the last year of his life conducted a jeweller and silversmith's business in Boston under the name of Rich & Willard.

Benjamin F. Willard married October 3, 1837, Emeline Maine.[3] She died January 20, 1892.[4] He died March 11, 1847.[5] Their daughter Ada Louisa, born 1839, died September 3, 1848. G. S.[6]

Benjamin F. Willard did not make very many clocks, and it is a matter of regret that he did not confine himself

[1] Roxbury Births. 1632 to 1844.
[2] Boston Directories. 1834 to 1838.
[3] Boston Marriages. 1800 to 1849.
[4] Boston Deaths for 1892. No. 853.
[5] Roxbury Deaths. 1633 to 1860.
[6] Grave Stone, Forest Hills Cemetery.

to that business, in which he no doubt would have become as renowned as his father. In 1844 he constructed at his brother's shop in Boston, an astronomical clock, which was one of the finest of its kind ever made. Every hole was jewelled with first-class sapphires, except the main arbor which had hardened steel. This jewelling was very costly. The plates were cast brass, very heavy, and the pillars massive. The pinions and all the steel work were miracles of polish, and the finish of the whole clock was wonderful. A very ingenious contrivance for lowering the pendulum by which by loosening a screw the pendulum rested on a massive support and allowed the movement to be taken away without trouble. The pendulum was mercurial, the mercury weighing fifty-six pounds. The movement was covered with a brass frame holding a plate glass whereby the back and sides could be seen from the outside. The mahogany case was also provided with plate glass windows. This mahogany case, a very fine one, was made specially to order by Charles C. Crehore.

For this clock (Plate 25) he was awarded a Gold Medal by the Massachusetts Charitable Mechanics Association. He also invented a revolving light for lighthouses, for which he received a patent from the United States Govern-  ment[7]. A revolving light of this design contracted for by

[7]Granted February 20, 1839. No. 1085.

B. F. WILLARD.

Signal Light.

No. 1,085.

Patented Feb. 20, 1839.

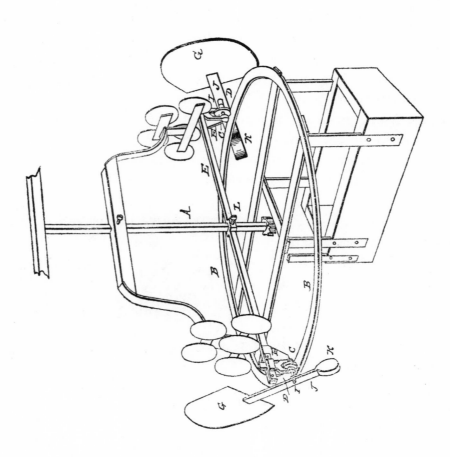

UNITED STATES PATENT OFFICE.

BENJAMIN F. WILLARD, OF BOSTON, MASSACHUSETTS.

IMPROVEMENT IN REVOLVING LIGHTS FOR LIGHT-HOUSES.

Specification forming part of Letters Patent No. **1,085,** dated February 20, 1839.

To all whom it may concern:

Be it known that I, BENJAMIN F. WILLARD, of the city of Boston, in the county of Suffolk and State of Massachusetts, have invented a new and useful Improvement in Revolving Lights of Light-Houses, called "Willard's Revolving Flashing Lights," which is described as follows, reference being had to the annexed drawings of the same, making part of this specification.

Owing to the multiplicity of the common revolving lights on our coast it has become very difficult for mariners to distinguish one light from another, so much so, indeed, that in some places it has become necessary to incur an additional expense of erecting two or three lights in the same place in order to render the light distinguishable from other lights. Now in order to save this additional expense and at the same time to render the light clearly distinguishable from all others is the object of this invention.

It is effected in the following manner: In addition to the ordinary clock-work heretofore used for imparting a regular rotary motion to the main or vertical shaft A, on which the lights are suspended, there is arranged upon and secured to the frame of the clock-work in a horizontal position a circular rim or railway B of any required diameter. Upon this horizontal stationary circular rim or railway B there is made to travel around in a circle upon its upper or flat side a vertical wheel C of a diameter adapted to the number of revolutions required to be performed in passing around upon said circular railway, whose axle D turns in suitable pulley-frames E, whose shank works loosely in an oblong mortise near the end of an arm F, extending horizontally from the vertical shaft beyond the periphery of the circular railway.

The object of having the frame of the wheel to rise or fall loosely in the mortise of the arm is to cause the wheel always to bear on the railway and be turned by the friction be-

tween their surfaces in contact and thus to turn the axle of the wheel, on one end of which is fixed a shade G, made of tin or any thin metallic substance, by means of a square socket I, fastened on the arms J of the shade and slipped over the end of the axle D, which is made square to fit said socket. The shade G is on one end of the arm and a weight K on the other to balance it. Said shade being then in a vertical position and directly in front of the lights, will when in motion cause the lights to appear and disappear in quick succession of sudden flashes. The other end of the arm is furnished with like wheel, axle, sliding frame, and revolving shade, operating and producing the same effect as those just described.

The arm for carrying around the wheels and to which they are attached is secured to the center vertical shaft, upon which the lights are suspended by a mortise in the center of said arm and a screw L, passing through the arm, by which it can be raised or lowered as a proper adjustment may require. It is made to extend each way in opposite directions and is placed directly under the revolving lights.

The invention claimed, and desired to be secured by Letters Patent, consists in—

The before-described method of rendering the revolving lights of light-houses distinguishable from other revolving lights by means of vertical revolving shades turned by wheels moving on a circular railway, to the axles of which the shades are fixed directly in front of the lights, which when in motion will cause the lights to appear and disappear in quick succession of sudden flashes, as herein set forth, whether produced by the combination of parts here described or any other combination substantially the same.

BENJAMIN F. WILLARD

Witnesses:
WM. P. ELLIOTT,
EDMUND MAHER.

the United States Government in 1828, was also built and tested at his brother's store, No. 9 Congress St. It was placed in the lighthouse at the entrance of Boston harbor about 1830 or 1834 and remained there many years until replaced by a modern Fresnel Light. The author gives an extract from a letter in regard to this light.[8]

"The Board states however that its records show that on 16 Oct. 1828, Gen. S. Pleasanton, Fifth Auditor of the Treasury and Commissioner of Revenue, then having charge of the Light-House Service, authorized Henry A. S. Dearborn, Esq., Superintendent of Lights for the District of Massachusetts, to procure an entirely new set of machinery for revolving Boston Light, and 'to accept the offer of Mr. Willard to supply it, on his improved plan, for two hundred and thirty dollars; employing him also to repair the old machinery.'

"The board regrets that owing to the inacessibility of its old records it is unable to give further information in reply to your inquiry."

Besides being a clock-maker and inventor, Benjamin F. Willard, was a remarkable penman. An example of his skill in this line is shown in Plate 27. This piece of work, largely reduced, unfortunately does not show all the details. The small circle in the lower right-hand corner has the Lord's Prayer written therein; in the original it has to be read with a magnifying glass.

Besides being a penman, he was an artist and painted many excellent pictures. His early death is to be regretted, as his accomplishments were such as would have made him eminent in any line of business, artistic or inventive, he might have undertaken.

[8]Letter from Thomas L. Casey, Lt. Col. Corps of Engineers, U. S. A. Lighthouse Board, Washington, D. C.

PLATE 27

PEN AND INK DRAWING, BY BENJAMIN F. WILLARD   SIZE OF ORIGINAL, 25¾ x 32¾ INCHES

PLATE 28

... Bailey, at Newbury Port.

Benjamin Willard

Clock Maker,

[illegible] informs the Publick, That [illegible] removed from *Lexington* to *Roxbury*, where those People [illegible] purchased Clocks of him and who depend on him for [illegible] Care of them, still may have the same Care taken by ap[illegible] to him at Roxbury, or by his Workman at Grafton, where [illegible] are made as well as at Roxbury: — Those that please [illegible] him with their Custom may have all sorts of com[illegible] House Clocks, and neatly Cased, at said Shop, [illegible] [illegible] than those that are purchased from foreign Countries. [illegible] hoped that this as well as other kind of Mechanical [illegible] will meet with suitable Encouragement here, as [illegible] have been large Sums of Money sent away for foreign Work [illegible] may be retained to the Emolument of this Country.

N. B. Also Watch Chains and Keys, Main Springs, Watch [illegible] and Chrystals, may be had at said Shop cheap for Cash.

BOSTON EVENING POST

DECEMBER 17, 1771

... boots, or any other Leath[illegible] requ[illegible] [illegible] Black.—Also, Marble-scented WASH BALLS.

Benjamin Willard

At his SHOP in Roxbury Street,

Performs the different Branches of Clock and Watch Work——And has for Sale.

Musical CLOCKS playing different Tunes, a new Tune every Day in the Week, and on Sunday's a Psalm Tune:—These Tunes perform every Hour without any Obstruction to the Motion or going of the Clock.—A new Invention for pricking Barrels to perform this Musick; and the Clocks are made much cheaper than any ever yet known. Also all sorts of common Spring Clocks and Eight Day Clocks are made at said Shop, and neatly cased and sold as cheap as can be Imported; and all warranted.

All the Branches of this Business is likewise carried on at his Shop in *Grafton*.

Imported from LONDON, in Captain Jenkins, and to ...

BOSTON GAZETTE

JULY 22, 1773

MUSICAL CLOCKS.

To be Sold Musical Clocks that go by Springs, also Musical Clocks that go by Weights, and play a different Tune each Day in the Week, on Sunday's a Psalm Tune. Inquire of BENJA. WILLARD Clock & Watch Maker in Roxbury Street near Boston. Where all sorts of Clocks are made in the newest Form, and Warranted to go without Variation, and many without Cleaning. Also Clock Cases are made at the same Place, in various Forms and in the best Manner, and Cheaper than can be purchas'd in London. Also Convey'd with Clocks to any Part of the Country.

All the Branches of this Business is likewise carried on at his Shop at Grafton.

TO BE SOLD,

BOSTON GAZETTE AND COUNTY JOURNAL

SEPTEMBER 5 AND OCTOBER 3, 1774

... Printing-Office, in Hanover-street, Boston, and by all the Printers in said Boston.

MUSICAL CLOCKS.

TO BE SOLD,

A NUMBER of Musical Clocks, which play a different Tune each Day in the Week, on Sunday a Psalm Tune. Enquire of

BENJAMIN WILLARD,

Clock and Watch Maker in Roxbury-street, near BOSTON.

Where all Sorts of Clocks are made in the newest Form and warranted to measure Time without Variation, and to go many Years without Cleansing. Also, Clock Cases are made at the same Place in various Forms, and in the best Manner, and cheaper than can be purchased in London, and conveyed with Clocks to any Part of the Country.

N. B. Said WILLARD likewise informs, that all Branches of this Business are carried on at his Shop at Grafton.

WILLIAM BRECK, begs leave to ...

MASSACHUSETTS SPY

OCTOBER 15, 1774

NEWSPAPER ADVERTISEMENTS

BENJAMIN WILLARD.

Benjamin[5] Willard, the second son of Benjamin[4] and Sarah (Brooks) Willard, was born in Grafton, March 19, 1743.[1] He was among the first of the early New England clock-makers, and the first of this famous clock-making family to engage in the business. It is a matter of regret that so little is known about him. Of his early life in Grafton, and of whom he learned his trade, the writer has been unable to find the slightest authentic information. He might have learned it from some journeyman clock-maker, or it might possibly have been the Morris referred to by Drake,[2] who was said to be the instructor of Simon Willard. The first authentic information we have of Benjamin Willard is found in two deeds,[3] where Benjamin Willard buys two lots of land with house, etc., in Grafton, of his father. These are dated May 18, 1764 and Aug. 20, 1766. These purchases would seem to indicate that he was intending to settle down in Grafton. He must have been clock-making before this date and had his factory at the old homestead. It will be noticed in his advertisements in the Boston papers he mentions his factory and workmen at Grafton. Between 1766 and 1771, he seems to have removed to Lexington, Mass., but the writer has been unable to find any trace of his stay in that town, where he lived, whether he had a workshop, or how long he remained. Probably his stay was not long. The author is of the opinion that Benjamin

[1]Original Grafton Records. Vol. i. Page 206.
[2]R. C. Vol. 34. Page 152.
[3]Worcester Deeds. Vol. 56. Page 13. Vol. 56. Page 15.

did not make clocks there wholly, but brought them down from Grafton, and put the finishing touches on at Lexington. Probably business was not good in Lexington, and deciding on a change, he announces his removal from Lexington to Roxbury, Mass., in the advertisement on Plate 28 which appears for the first time in the *Boston Evening Post*, December 17, 1771. This advertisement appears at intervals in the *Evening Post* until February, 1772.

Equally barren of information is his stay in Roxbury. The writer has tried to ascertain exactly where he lived, but has only succeeded in locating him in Roxbury Street, for in his advertisement in the *Boston Gazette*, February 22, 1773, he speaks of his shop in Roxbury Street. In his advertisement of Sept. 5 and Oct. 3, 1774, in the *Boston Gazette* and *County Journal*, he also speaks of his Roxbury Street shop.

As his brother Simon Willard's shop was only a few hundred feet from the Boston line, the question sometimes has been raised as to whether Simon might not have bought Benjamin's business out. The last trace the writer has of Benjamin Willard's stay in Roxbury is in the Roxbury Tax Lists,[4] where the entry is found of Benjamin Willard, 1 Poll, Real Estate £12, Pers. Estate £5. As he is assessed for only one poll it would indicate he had no partner or apprentices. His last advertisement occurs in the *Massachusetts Spy*, October 15, 1774. All his advertisements are noticeable for their quaintness.

It is quite likely that he did not spend all his time in Roxbury, but passed the winter months in Grafton. At the outbreak of the Revolution he probably returned to Grafton.

Roxbury Tax Lists. Vol. 5, 1774. Westerly Parishes of Roxbury.

The author has no information to show that Benjamin served in the Army, as did most of his brothers, and there is a gap between 1775 and 1783, where all trace of him is lost. In 1783 he reappears in Grafton and in legal difficulties with one Daniel Willard.[5] Also in 1783, the birth of one of his children is recorded in the Grafton records, showing that during this interval he must have returned to his native town and perhaps was married there. In 1788 he appears buying an estate in Grafton of Cyrus French.[6] There is no record of his marriage that the writer can find. He perhaps married in Grafton and his wife might have been a native of that place, but there is no information to show where she came from.

> Benjamin Willard married Peggy [Margaret] Moore.[7] She died in Grafton, June 1, 1837, ae 85.[8] He died in Baltimore, Md., September, 1803.[9]

Children of Benjamin and Margaret Willard.

> Elizabeth Moore, born May 12, 1783.[10]
> Margaret, born May 23, 1785.[10]
> Benjamin, born Aug. 6, 1787.[10] Died Jan. 5, 1801; ae 13.[8]
> Nancy M., born about 1793. Died June 21, 1816; ae 23.[8]
> Martha—————

The births of the daughters Nancy M. and Martha are not recorded in the Grafton Records, but Martha is mentioned in the will of her mother.[11] Benjamin seems to have remained in Grafton with his family for some time.

[5]Worcester Deeds. Vol. 90. Page 126.
[6]Worcester Deeds. Vol. 103. Page 648.
[7]Note in manuscript of Joseph Willard, author of the Willard Memoir.
[8]Grave Stone. Old Grafton Cemetery.
[9]*Worcester Spy*. Date of October 5, 1803.
[10]Original Grafton Records. Vol. 1. Births and Deaths. Page 146.
[11]Worcester Probate. 66002.

In 1790 he is recorded as living in Worcester Town.[12] In 1791 it looks as if he were contemplating a change, for he sells land in Grafton to Thomas Axtell, Jr.,[13] and after this there is nothing authentic about him until 1798, when Benjamin Willard seems to have fallen on evil days. He appears to have got into a lawsuit or some legal tangle with one John Taylor of Northboro, Mass., who got judgment against him for $1901.94 and costs.[14] This lawsuit not only seems to have ruined him, but landed him in jail besides, (probably for debt). After this, information as to his movements is very vague and unsatisfactory. That he remained in Grafton some time after his lawsuit is proved by the United States Census return for 1800, in which he and his whole family are recorded.

WORCESTER COUNTY, MASSACHUSETTS, 1800.

NAMES OF TOWNS	NAMES OF HEADS OF FAMILIES	FREE WHITE MALES					FREE WHITE FEMALES				
		Under 10 years of age	Of 10 and under 16	Of 16 and under 26, including heads of families	Of 26 and under 45, including heads of families	Of 45 and upwards, including heads of families	Under 10 years of age	Of 10 and under 16	Of 16 and under 26, including heads of families	Of 26 and under 45, including heads of .amilies	Of 45 and upwards, including heads of families
Grafton	Benjamin Willard	—	1	—	—	2	2	1	1	1	—

At some time in 1801, he must have gone to Boston, but all efforts to locate him have failed. His name does not appear in the Boston Tax Lists, or in the Directory. At some time in 1803 he went to Baltimore, Md., where he died, for in

[12]Worcester Town. First Census of the U. S. 1790. Page 245.
[13]Worcester Deeds. Vol. 111, Page 430, and Vol 113, Page 647.
[14]Worcester Deeds. Vol. 133. Pp. 466-470.

PLATE 29

BENJAMIN WILLARD

HALL CLOCK

OWNED BY

ARTHUR W. WELLINGTON

BOSTON, MASS.

AARON WILLARD

HALF OR SHELF CLOCK

OWNED BY

DWIGHT M. PROUTY

CHESTNUT HILL, MASS.

PLATE 30

OLD BENJAMIN WILLARD HOMESTEAD, GRAFTON, MASS.

the issue of the *Worcester Spy,* for October 5, 1803, is the following death notice: "At Baltimore, Mr. Benjamin Willard, formerly of Grafton, where he has left a wife and four children."
Also in the *Columbian Sentinel,* Boston, September 28, 1803:
"Deaths. At Baltimore, Mr. Benjamin Willard of this
Town, where he has left a wife and four children." Probably he went to Baltimore to try and establish the clock-
making business there. Benjamin Willard left no will, and
May 1, 1804, his widow is appointed administratrix of his
estate.[15] Benjamin Willard left very little property, the
appraisers only reporting personal property, value of $75.09,
no real estate.[15] His lawsuit evidently had stripped him of
everything. Of Benjamin Willard's clock-making, his factory, or his methods of working, the author has no information whatever. Histories of Grafton barely mention the
name.

He did not make a very good clock compared with
his brother Simon, and apparently made only the tall Hall
clock. The writer has never heard of his making the
Half or Shelf clock, nor has he ever seen one. Specimens of his Hall Clocks marked Grafton, Lexington and
Roxbury have been examined by the author, but he has
never seen one marked Boston or any other place other
than the above. Plate 29 shows by far the finest example
of Benjamin Willard's clocks that the author has ever seen.
Its condition is almost perfect, having all the original
parts, even to the wooden pendulum rod. The case is
a very handsome one of selected mahogany. The dial
is an imported one of heavy engraved brass, giving the

[15]Worcester Probate. 65849.

changes of the moon and the days of the month. At
the top of the dial is the quaint inscription in a running
hand, "The man is yet unborn that duty weighs one
hour." The centre of the dial has the words "Benjamin
Willard, Roxbury." The date of this clock would probably be between 1771 and 1775. It would be interesting
to know if Benjamin Willard ever made any clocks in
Baltimore. All the clocks seen have had handsome brass
dials evidently imported from England. All the early
New England clock-makers were obliged to import their
dials from England, and would finish them up at their
own shops. Benjamin Willard was a great wanderer and
doubtless was in other places besides the ones where he
has been located. His family is extinct in the male line.
The old Benjamin Willard house (Plate 30) is still standing in Grafton, Mass., although somewhat changed. The
place is about three miles from Grafton Center, in the Farms
District, on the crossroad to Westboro. The ell is a later
addition and the main dwelling house has had an additional
story added. The front door is the original one, and the
interior still retains the original stairs, and the rooms are
very little altered. On the old stone step are carved the
initials B. W. (Plate 31.)

The author is of the opinion that this house was
built by Benjamin[2] Willard, who in his will[16] drawn
March 4, 1717-18, probated Aug. 18, 1732, speaks of his
"homestat in Hassanimisco" [Grafton]. This being the
case the old homestead would certainly date from 1717
and perhaps earlier.

[16]Worcester Probate. 65847.

PLATE 31

PORTION OF THE OLD DOOR AND DOORSTEPS ON THE
BENJAMIN WILLARD HOUSE AT GRAFTON, MASS.

PLATE 32

AARON WILLARD

SHELF OR HALF CLOCKS

OWNED BY

HOWARD MARSTON

BOSTON

AARON WILLARD.

Aaron[5] Willard, ninth son of Benjamin[4] and Sarah (Brooks) Willard, was born in Grafton, Mass., October 13, 1757.[1] Of his early life in Grafton, and to whom he was apprenticed, the author has no information whatever. An article in a recent paper states that Aaron Willard learned his trade of Alexander T. Willard and Philander J. Willard, clock-makers, of Ashby, Mass. As they were born in 1772 and 1774 respectively, and did not learn their trade until about 1798, the improbability of this statement is evident, for in 1798, Aaron Willard was well established in his business in Boston. The author is very much inclined to the opinion that Aaron Willard learned his trade of one of his brothers, either Benjamin or Simon. The first authentic information of Aaron Willard is in 1775, when in response to the alarm of April 19, he marched from Grafton to Roxbury. His military record is as follows:[2]

"Aaron Willard, Grafton. — Private, Capt Aaron Kimball's co of Militia [Col] Artemas Ward's regt, which marched in response to the alarm of April 19, 1775, said Willard marched April 19, 1775, service 1 week, reported enlisted into the army April 26, 1775, *also* Capt Luke Duruy's Co, Col Jonathan Ward's regt, muster roll dated Aug. 1, 1775, enlisted April 26, 1775, service 1 mo. 10 days, reported enlisted into the train, June 3, 1775, also company return, probably Oct. 1775."

Drake[3] gives an anecdote about Aaron Willard's services at this period. He served longer in the army than any of

[1] Original Grafton Records. Vol. 1. Births. Page 206.
[2] Massachusetts Soldiers and Sailors of the Revolution. Vol. 17. Page 379.
[3] R. C. Vol. 34. Page 374.

his brothers. This seems to have been the extent of his army service, and after his discharge he probably returned to Grafton, where he doubtless finished his apprenticeship. In 1780 he came to Roxbury at the same time as his brother Simon. His name appears in 1783 in the Roxbury Tax Lists, assessed for 2 polls, 3£ Real Estate, 6£ Personal.[4] This would seem to show that he had an apprentice or workman with him. Drake[5] says that Aaron Willard first kept where the apothecary shop numbered 2224 Washington St. now is. This would place him some little distance above his brother's, Simon Willard's, shop, now No. 2196.

Aaron Willard married 1st, Catherine Gates of Roxbury, March 6, 1783.[6] She died July 30, 1785, ae 22.[7]

Children of Aaron and Catherine (Gates) Willard.

> Aaron, born June 29, 1783.[6]
> Nancy, born July 14, 1785.[8]

Aaron Willard married 2nd, Polly Partridge, Nov. 1, 1789.[9] She died Oct. 10, 1846, ae 85.[7] He died May 20, 1844, ae 87.[7]

Children of Aaron and Polly [Mary] (Partridge) Willard.

> Polly, born Dec. 1790.[8]
> Sophia and Emily (gem), born Nov. 27, 1792.[8]
> Catherine Gates, born Oct. 15, 1794.[10]
> Jane J., born about 1798. Birth not recorded. Died Aug. 13, 1886.[11]
> Charles, born July 12, 1800.[12]

[4]Roxbury Tax Lists, for November, 1783.
[5]R. C. Vol. 34. Page 153.
[6]Roxbury B. M. and D. 1630 to 1785. Manuscript Copy. Page 174.
[6]Ibid. Page 114.
[7]Grave Stone. Eustis Street Cemetery, Roxbury, Mass.
[8]Roxbury Births. 1633 to 1844.
[9]R. C. Vol. 30. Page 112.
[10]Date of birth given from family records of Mr. Watson Gore, Braintree, Mass.
[11]Lexington, Mass. Vital Records. Page 482.
[12]Boston Births. 1800 to 1849.

Henry, born May 1, 1802.[12]
Morris, born Oct. 21, 1808.[12]

Aaron Willard's children survived him with the exception of Charles, who probably died young. The author can find no record of him, and he is not mentioned in Aaron Willard's will. Nine children are mentioned.[13] Most of his children moved away, going to New Orleans and Philadelphia, Henry and Jane J. Willard remaining in Boston.[13]

It may be noted here as a rather curious fact that Aaron Willard, Elnathan Taber, Abel Hutchins, and Samuel Curtis, all clock-makers, married sisters, daughters of Thaddeus Partridge of Boston and Roxbury.[14] Another daughter, Abigail Partridge, married John Pierce Sawin of Roxbury, Dec. 16, 1798.[15] Their son, John Sawin, born Sept. 13, 1799,[16] was apprenticed to Aaron Willard, Sr. After serving his apprenticeship, he set up in the clock-making business in Boston, his place of business being at various places, but principally on Court St.[17] He was frequently employed by Simon Willard, Jr., & Son to make clocks for them. He made a very good clock. He died in 1863.[18] Aaron Willard was a good, keen business man, and after settling down in Roxbury, saw that only a bare living was to be made in making clocks himself, and peddling them around the country single-handed, and the only way to make money was to manufacture clocks

[12]Boston Births. 1800 to 1849.
[13]Suffolk Probate. Vol. 142. Pp. 369 to 371.
[14]Norfolk Probate. 14132.
[15]Roxbury Marriages. 1632 to 1860.
[16]Roxbury Births. 1632 to 1844.
[17]Boston Directories. 1848 to 1862.
[18]Boston Deaths for 1863. No. 1008.

on a large scale. He therefore decided to start a factory and make clocks wholesale, and of a cheaper grade.
Whether he had a factory in Roxbury first, the author
is inclined to doubt; at least no evidence can be found
to indicate that he had.

The evidence gathered seems to indicate that Aaron
Willard established his factory in Boston, on the Neck,
a little distance from the Roxbury line, not taking up his
residence there until some time after. Just when he
moved to Boston is not quite certain. In 1788 he calls
himself a resident of Roxbury.[19] In 1792 he bought an
estate in Boston,[20] on the Neck, which probably is the
one he made his residence, and was afterwards numbered
843 Washington St.,[21] and where he lived until he died.
In this deed also he calls himself of Roxbury. His name
does not appear in the United States Census of 1790 for
Boston, or in the Boston Directory for 1796, but is found
resident in Roxbury in 1790.[22] His name and residence
is given in the Boston Directory for 1798, as "Aaron
Willard, Clock-maker, on the Neck." His place is described[23] as "Aaron Willard, owner and occupier of wooden
dwelling, East by Washington, South by land of Sam'l
and Arnold Welles, North by William Fisk, wood house
480 square ft. Wood Barn 792 square ft. wood, Land
30,000 square ft., house 1464 square ft. 3 stories, 36 windows, value $3000."

It would seem therefore from the evidence shown that

[19]Suffolk Deeds. Vol. 163. Pp. 85-86.
[20]Suffolk Deeds. Vol. 173. Pp. 220-221.
[21]Boston Directories. 1825 to 1848.
[22]First Census of the U. S. for Mass., 1790. Roxbury Town. Name not indexed.
[23]R. C. Vol. 22. Page 442.

his factory was first established in Boston, and that he took up his residence there some time between 1792 and 1798. His factory was in the rear of his house, which fronted on Washington St., and was connected with the house by a

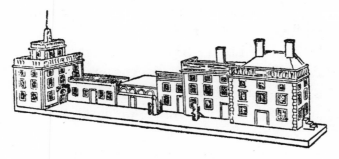

MODEL OF THE WILLARD CLOCK FACTORY.

row of buildings, fronting on a lane that led off Washington St. This lane is now probably Derby Place. These buildings on the lane were the storehouses of Aaron Willard, and the workshops of the various trades who probably rented them of Aaron Willard. Among these were Charles Bullard,[24] ornamental painter, Henry Willard, clock-case maker, Pratt & Walker, cabinet makers, often described in the Directories as being in rear of 843 Washington St.

Aaron Willard's house and factory were landmarks in this section of the town. In course of time a little colony of clock-makers and the various trades dependent on them established themselves in the vicinity. Taking the Roxbury line in Washington St. as a centre, in 1816 and later, a radius of half a mile would include nearly all the clock-makers of note, with the attendant trades. Simon Willard, William Cummens, Elnathan Taber, clock-

[24]Boston Directories. 1816 to 1844.

makers, Nehemiah Munroe, cabinetmaker, on the Rox-
bury side, Aaron Willard, Aaron Willard, Jr., clock-makers,
Charles Bullard, John R. Penniman, Samuel Washburn,
John Green, Jr., painters, William Fisk, Pratt & Walker,
Thomas Bacon, Spencer Thomas, cabinetmakers, Lewis
Lauriat, goldbeater, Nolen and Curtis, dial makers, Simeon
Gilson, William Abbot, brass founders, Thomas Ayling,
turner, and Thomas Wightman, carver. Besides these, the
lead works and mahogany mills were close by. All these
supplied the clock-makers more or less.

Just why Aaron Willard selected this part of Boston
instead of going to the city proper is hard to say. This part
of Boston was for a long time almost inaccessible, and considered completely out of the way. Drake[25] says, "In season
of full tides portions of the Neck were covered with water,
rendering it almost impassible in the spring . . . The appearance of this avenue sixty years ago was desolate and forbid-
ding enough. Between Dover St. and the Roxbury Line
there were but eighteen buildings in 1794."

Aaron Willard prospered and he did a large business,
employing as high as twenty or thirty workmen in his later
years. There is not the slightest evidence the author can
find to show that Aaron Willard's brothers, Benjamin, Simon
or Ephraim, were ever associated with him. Aaron Willard
never seemed to advertise his business. The author, after a
careful search, has never been able to find an advertisement of
any kind in the newspapers. It was not the custom of those
days to advertise much. Aaron Willard made the tall Hall
Clock, Half or Shelf Clock, Timepiece, Gallery Clock, and

[25]R. C.   Vol. 34.   Pp. 66-67.

Regulator Clock. When he first started in business he confined himself to the Hall Clocks and Half Clocks. The Half Clock he made in great quantities and in the greatest variety of styles. Plate 29 is a fair example of the solid mahogany cases, and Plate 32 is an exceptionally good example of the glass front style, the glass being especially good. In Plate 32 the ground color of the glass is a beautiful sea green tint. The early specimens of the Half Clock are rather clumsily made, especially the weights. He continued the style of clock that Simon Willard abandoned in 1780. Many of the cases of these Half Clocks are beautifully inlaid. He made the Half Clock with solid mahogany cases, until Simon Willard introduced the glass front style, which Aaron Willard promptly copied. Sometimes the cases are found made of yellow mahogany, oak or cherry. Many of the cases of his Hall Clocks are also beautifully inlaid. Like his brother, Simon Willard, he made what was called the Ship Clock, a tall clock having the figure of a ship on the dial, secured to the pallets, and rocked by the swing of the pendulum.

After Simon Willard brought out his Timepiece of 1802, Aaron Willard gradually abandoned the manufacture of the Half Clock, and made the Timepiece in increasing quantities. Being desirous of making a cheaper grade of clock, his glass fronts were never as elaborate as those of Simon Willard, unless it was made on an order. Plate 20 is a very fair type of Aaron Willard's Timepiece. The design on the door glass is one much used by him, also the narrow, glass front design. The mahogany case is decorated with a narrow beading, gilded, and is also provided with a base piece or bracket. The top ornament is original. Aaron Willard generally used a gilded acorn as a top ornament, but sometimes used the ball and spike, rarely the spread eagle. He almost invariably put his name on the dials of his clocks, as did his son Aaron, Jr. Aaron Willard made the striking Timepiece also. He did not make many of them, as they were expensive.

A style of clock affected by Aaron Willard, although he does not seem to have made many of them, was a Regulator Clock, having a concave or dished dial. He made these with either solid mahogany cases or glass fronts. An example of the solid mahogany case style is shown in Plate 33. An excellent example of the glass front style may be seen at Crosby's Restaurant on School St., Boston. This style of clock was also made by Aaron Willard, Jr. Aaron Willard also made the Church or Turret Clock, but the author is inclined to think he did not make very many of them.

As may be imagined in conducting a business so many years on a large scale, the number of clocks turned out by the Aaron Willards was very large, and consequently their clocks are by no means uncommon, being found practically over the whole of New England. In his early clock-making days Aaron Willard used to peddle clocks around the country, his beat being the South Shore. The clocks made by Aaron Willard and Aaron Willard, Jr., vary considerably in quality, being seemingly dependent on the ability of the workmen they employed. Sometimes the clock movements are especially good, and again a poor quality of work will be noted. As a rule the men employed by Aaron Willard,

Jr., did not seem to be such good workmen, owing perhaps to the gradual decline of the apprenticeship system.

About 1823 or a little later Aaron Willard, having made a very comfortable fortune for those days, decided to retire from active business, and his son, Aaron, Jr., took charge of the whole business, making but few changes in the style of clocks turned out, which changes are noted in the life of Aaron Willard, Jr. Aaron Willard died in 1844. His will mentions his house 843 Washington St., and his property was divided among his nine surviving children.[26] None of his sons, with the exception of Aaron, Jr., and Henry, seem to have had anything to do with the clock-making business. After Aaron Willard's death, the little clock-making colony gradually drifted away and finally in 1850 [27] his place was sold, and a business that had existed for over sixty years became a thing of the past.

[26]Suffolk Probate. Vol. 142. Pp. 369-370-371.
[27]*Boston Advertiser.* December 16, 1850.

AARON WILLARD, JR.

Aaron[6] Willard, Jr., only son of Aaron[5] and his first wife, Catherine (Gates) Willard, was born in Roxbury, Mass., June 29, 1783.[1] He learned his trade in his father's clock factory, and he lived most of his life in Boston, on the Neck, engaged in the clock-making business. After serving his apprenticeship he was for a short time in partnership with a Spencer Nolen, advertising themselves as Willard & Nolen, clock and sign painters, Boston Neck.[2] Spencer Nolen afterwards married Nancy Willard,[3] Aaron Willard, Jr's., sister. The partnership did not last long, for in 1809, Aaron Willard, Jr., is recorded as a clock-maker on Washington St.[4] He probably was employed at his father's factory. In 1815 he bought an estate on Washington St., Boston Neck.[5] Here with some additions to his place, he lived for over thirty years. His place, numbered 815[6] Washington St., was where the Washington Market now stands.

He was very fond of flowers and gardening and his place was laid out with very beautiful flower beds, and his house had a large greenhouse attached to it. An orchard and vegetable garden were in the rear of the house. A stone post with a small dial clock on the top stood in front of the house. There was no clock factory

[1]Roxbury B. M. and D. 1630 to 1785. Manuscript copy. Page 114.
[2]Boston Directory. 1806.
[3]R. C. Vol. 30. Page 264.
[4]Boston Directory. 1809.
[5]Suffolk Deeds. Vol. 247. Pages 280-281.
[6]Boston Directories. 1825 to 1848.

PLATE 33

AARON WILLARD

REGULATOR CLOCK

OWNED BY

THE AUTHOR

AARON WILLARD, Jr.

LYRE CLOCK

OWNED BY

PATRICK MEANIX

on the place that the author knows of. Z. A. Willard says he often had occasion to go past there, and has no recollection of any shop or factory on the premises, unless the clock-making was carried on in the house. As Aaron Willard, Jr., did a large business and employed many men, it is more than likely the clock-making was done at his father's factory, especially as he took over his father's business about 1823.

Aaron Willard, Jr., married, Jan. 7, 1816, Ann Dorr.[7] She died, June 7, 1842, ae 61.[8]

Children of Aaron, Jr., and Ann Willard.

> Emily, born Dec. 27, 1816.[9]
> Anthony Mayben, born July 11, 1819.[9]

Aaron Willard, Jr., married as his second wife, Nov. 10, 1855, Emeline Davenport,[10] who survived him. He died in Newton, Mass., May 2, 1864.[11] There were no children by the second marriage. The births of Aaron Willard, Jr.'s, children are not recorded in the Boston Records, but are found in the Records of the Hollis St. Church, a copy of which is in the City Register's office. As it is only a manuscript copy, the author simply gives the record for what it is worth, not having seen the original records. The son evidently died young as he is not mentioned in his father's will. The daughter, Emily, did not survive her father.

[7] Boston Marriages. 1800 to 1849. Page 418.
[8] Boston Deaths. 1810 to 1848.
[9] Hollis St. Church Records. Baptisms. Manuscript Copy. Page 126.
[9] Ibid. Page 134.
[10] Massachusetts Vital Records. State House, Marriages 1855.
[11] Middlesex Probate 37383.

Having amassed quite a comfortable fortune, Aaron Willard, Jr., decided to give up his business and retire. He began to sell his estate [12] and gradually closed out his business. Having sold his place at No. 815 Washington St., he lived for a year or so at his father's old place at No. 843 Washington St.[13] He purchased an estate in Newton,[14] Mass., and removed to that place about 1850, where he lived until his death. Whether he made any clocks in Newton, the author is unable to say, but it is more than probable he did not. In some deeds [15] he calls himself gentleman, no occupation mentioned. As he was passionately fond of gardening he probably amused himself by farming. His place in Newton was what is now called Oak Hill.[16]

Aaron Willard, Jr., made the Timepiece, Hall clock, Regulator, and Gallery clock, and conducted a general repair business. Timepieces with his name on the dial do not appear until 1823, at which time Aaron Willard, Jr., took over his father's business. He made the Timepiece in great quantities. He often made the bezel case of wood, something that Simon Willard never did. Aaron, Jr., did not have as fine front glasses painted for his Timepieces as did Simon Willard, as he was making a cheaper grade of clock. He, however, was inclined to use a little gilding on his Timepieces, and commonly added a base-piece, something Simon Willard never did, except on his Gift Clocks. Aaron Willard, Sr. and Jr., imitated the Gift Clock

---

[12]Suffolk Deeds. Vol. 545. Page 263.
[13]Boston Directory. 1848-49.
[14]Middlesex Deeds. Vol. 613. Page 474.
[15]Middlesex Deeds. Vol. 608. Page 613. Vol. 684. Page 277.
[16]Middlesex Probate. 37383.

in a way, but never had the beautiful front glasses or the white enamelling.

One variety of Timepiece, original with Aaron Willard, Jr., was the so-called Lyre clock, of which Plate 33 is a good example. This type has been erroneously attributed to Simon Willard and Aaron Willard, Sr. It was peculiar to Aaron, Jr., and he made them in great variety and quantity after he took over his father's business. Some had solid mahogany cases, some carved mahogany, as in Plate 33, and others had painted glass fronts. This style of clock was occasionally imitated by later clock-makers. Aaron Willard, Jr., employed a good many workmen, and turned out large numbers of clocks. The author has seen a Timepiece of his numbered 3482. He made the Church or Turret clock, although it is probable he did not make very many of them. The author has seen only one. This is in the tower of the West Church (Evangelical Congregational), Grafton, Mass.

HENRY WILLARD.

Henry[6] Willard, son of Aaron[5] and his second wife, Mary (Partridge) Willard, was born in Boston, May 1, 1802.[1] Birth was also recorded in Roxbury.[2] He married, Oct. 16, 1831, Frances A. Williams.[3] There seems to have been only one child born of this marriage, Mary A. E. H. Willard, birth not recorded, but in 1863 this name is recorded as being the daughter of Henry and Frances A. Willard, age 24.[4] Henry Willard is said to have married a second time, but the author can find no record of the second marriage, or the deaths of the first or second wives.

Henry Willard was an apprentice of William Fisk, a noted cabinet-maker, who had a shop near Aaron Willard. He lived for the greater part of his life at No. 843 Washington St., Boston Neck, working at his trade which was cabinet-making, his specialty being clock-case making.[5] In 1847, Henry Willard removed to Canton,[6] Mass., making clock-cases for Simon Willard, Jr., & Son, and also running a farm, which was directly under the shadow of Great Blue Hill.[7] In 1887, he went to Boston, where he died the same year.[8] Henry Willard was a notable cabinet-maker and did very fine work. He made clock-cases for his father, Aaron Willard, and his brother, Aaron Willard, Jr., William Cum-

[1]Boston Births. 1800 to 1849. Page 306.
[2]Roxbury Births. 1632 to 1844.
[3]Roxbury Marriages. 1632 to 1860.
[4]Boston Marriages for 1863. No. 1262.
[5]Boston Directories. 1825 to 1847.
[6]Norfolk Deeds. Vol. 172. Page 20.
[7]Norfolk Deeds. Vol. 584. Page 361.
[8]Boston Deaths for 1887. No. 7782.

mens, Elnathan Taber, Simon Willard, Jr., & Son, but he did not make them for Simon Willard. He made a model of his father's place on Boston Neck (Page 89). This was for a long time perched on a pole on Henry Willard's place in Canton, Mass. This model is probably owned by the Hemmenway Estate, which bought the place.[9]

[9]Norfolk Deeds. Vol. 595. Pages 635-636.

EPHRAIM WILLARD.

Ephraim[5] Willard, eighth son of Benjamin[4] and Sarah (Brooks) Willard, was born in Grafton, Mass., March 18, 1755.[1] Nothing is known of his early life in Grafton, except that he was there in 1775, when he marched in response to the Lexington alarm. His record is as follows:[2]

> "Ephraim Willard, Grafton, Private. Capt. Aaron" "Kimball's co of Militia [Col] Artemas Ward's regt, which" "marched in response to the alarm of April 19, 1775, said" "Willard marched April 19, 1775, discharged April 29." "service 1 week, 5 days, reported returned home."

The first definite information as to his residence and occupation is in 1777. At this date he appears in Medford,[3] Mass., as a clock and watch-maker, and in company with a William Gowen of Medford, Mass., goldsmith, buys a tract of land in Grafton, Mass., of Daniel Willard, and in 1778, sells the same to an Ebenezer Hall,[4] Jr., of Medford. In each of these transactions his occupation is given as a clock-maker, and his residence in Medford, and no wife mentioned. He appears to be in partnership with this William Gowen. The writer can find no mention of Ephraim Willard in any of the Medford Records or Histories. How long he remained in Medford is not known, but in 1798, he is recorded as a resident of Roxbury,[5] Mass. He might have been married about this time for the birth of a son to

[1]Original Grafton Records. Births. Vol. 1. Page 206.
[2]Massachusetts Soldiers and Sailors of the Revolution. Vol. 17. Page 383.
[3]Worcester Deeds. Vol. 78. Page 255.
[4]Worcester Deeds. Vol. 80. Page 130.
[5]Worcester Deeds. Vol. 141. Pp. 288-289.

PLATE 34

EPHRAIM WILLARD

THEODORE W. GORE

AUBURNDALE, MASS.

HALL CLOCKS OWNED BY

WILLIAM CUMMENS

HENRY CLAP KENDALL

DORCHESTER, MASS.

Ephraim and Hepzibah Willard, June 24, 1779, is recorded in Roxbury,[6] but the author can find no record of the marriage.

About 1801, he removes to Boston, and is recorded as buying in Sheafs Lane, now West St., but in the deeds calls himself a merchant.[7] In the Assessors' Lists, however, his occupation is given as clock-maker,[8] and he is assessed for 2 polls, and $300 real estate. In a deed dated Jan. 6, 1801, his wife's name appears for the first time signing herself as Hepsee Willard.[9] After this his occupation is given as merchant, although he might have carried on the clock-making business as a side line, as in 1803 he is spoken of in a certain transaction as clock-maker, otherwise trader.[10]

In 1805 appears the last record of Ephraim Willard's stay in Boston,[11] and all positive knowledge of him is lost except that he is known to have removed to New York City. The author had an opportunity recently to look over the manuscript of Joseph Willard, author of the Willard Memoir, and Ephraim Willard was noted as having moved to New York. The author has obtained very little satisfaction in his attempts to obtain information about Ephraim Willard in New York. Inquiry at the Department of Health gave no information, and no record of will could be found at the Surrogate's office.

Search of the New York City directories gives the name of an Ephraim Willard, shipmaster, living at 55 Elizabeth St., from 1811 to as late as 1815-6, but his name does not

[6]Roxbury Births. 1632 to 1844.
[7]Suffolk Deeds. Vol. 196. Page 179.
[8]Assessors' List of Ward 12. May, 1801. Page 5.
[9]Suffolk Deeds. Vol. 196. Page 178.
[10]Suffolk Deeds. Vol. 205. Page 119.
[11]Suffolk Deeds. Vol. 210. Page 267.

appear for the next few years. In the 1825-6 directory, the
same name appears, giving his occupation as watch-maker,
and the address as 56½ Bowery, and the name appears with
variations of address until the 1832-3 directory, at which
time he lived at 180 Forsyth St. As it does not seem probable
that a ship-master would become a watch-maker, the second
Ephraim Willard may have been a son of the former, or he
may have been the Ephraim Willard from Boston, or his son.
After 1833, the name disappears. He may have removed to
some other locality, which in view of his inclination to
wander, is not unlikely. As the author was unable to make
a personal search of the New York City Records, he was
obliged to let the matter rest here, hoping at some future
time more information might be obtained.

Of Ephraim Willard's early life in Grafton, whom he
was apprenticed to, or if he was in business with any of his
brothers, the author has no knowledge whatever. In fact,
the most diligent search has failed to throw any light on
the early life of any of the brothers. It is more than likely,
however, that he learned his trade of either Benjamin or
Simon. That there was a fourth brother in the business in
this clock-making family was a great surprise to the author.
It was not until the discovery of a quit-claim deed where
nearly all the family with their residences and occupations
were given,[12] that the fact became known. The author's
father, Z. A. Willard, states that he never heard of Ephraim
Willard, and never heard his father or grandfather mention
him, in any way, and also he had never seen or heard of any
clocks made by him.

[12]Worcester Deeds. Vol. 141. Pp. 288-289.

Although Ephraim Willard appears to have been in the clock-making business for over twenty-years, clocks made by him seem to be very rare. The author has seen only one. This clock (Plate 34) from information given the author seems to have been made perhaps between 1780 and 1795.[13] It is in very good condition, made with a seconds hand, and has a nicely decorated dial; the case is of mahogany, inlaid; the brass top ornaments are probably original. Judging from this one clock, Ephraim Willard appears to have made a fair clock. The author has no knowledge that Ephraim Willard ever made the Half Clock or the Timepiece. He evidently made only the Hall Clock. He seems to have been as great a wanderer as his brother Benjamin[5] Willard.

[13]Genealogical notes given by Mr. Theodore W. Gore, Auburndale, Mass.

LEVI AND ABEL HUTCHINS.

Simon Willard had many apprentices; for his great reputation naturally drew many young men to his workshop. The author has endeavored to ascertain the names of the various apprentices of Simon Willard, but has only succeeded in finding five about which there is no doubt. There were others, but nothing authentic can be found about them. The two brothers, Levi and Abel Hutchins, children of Gordon and Dorothy (Stone) Hutchins, were born in Harvard, Mass.[1]

Levi, August 16, 1761.[1]
Abel, March 16, 1763.[1]

Levi Hutchins says,[2]

"My brother and I, entertaining a desire to learn the same trade, commenced our apprenticeship at nearly the same time, [1783]. The name of the ingenious man of whom we learned this business was Simon Willard of Roxbury, Mass. One trait in his character was, I think, caution. For instance, during a thunder storm he was particular in requiring his men and apprentices to suspend work in the Shop, believing there was much danger to be apprehended on such occasions from the action of electricity when using files and metallic tools. After three years apprenticeship under Mr. Willard, I went to Abington, Ct., where I served eight months to acquire some knowledge of the art of repairing watches.

Shortly afterwards I returned to Concord, hired a Shop on Main St., purchased materials and established the business of brass clock-making, no person having before undertaken this enterprise in New Hampshire. Soon

[1]History of the Town of Harvard, Mass. 1732-1893. Henry S. Nourse, A. M. Page 555.
[2]Autobiography of Levi Hutchins, by his youngest son, privately printed, Cambridge, 1865. Page 53, 56.

after his marriage in 1788, my brother Abel became my partner in the clock-making business, and our shop stood a little in the rear of a large well finished dwelling house, three stories high, which we jointly purchased and occupied, with our families; situated in the central part of the Main Village on the eastern side of the road or Street.   Peabody Atkinson and Jesse Smith were two of our apprentices, my brother and I were successful in our business.   We carried on clock-making together twenty-one years.   Our names may now be seen on the faces of many timekeepers and probably there are eight-day clocks or timepieces of our manufacture in all the original thirteen States of the Union, two eight-day clocks we made to order, and sent to the West Indies."

Levi Hutchins died June 13, 1855.   Ae 93 years, 9 months, 26 days.[4]
Abel Hutchins died April 4, 1853.   Ae 90 years, 19 days.[5]

The Hutchins made the tall Hall clock, and the Half or Shelf clock and undoubtedly made the Timepiece, although the author has never seen one bearing the Hutchins' stamp. Specimens of the tall Hall clock made by the Hutchins are quite common.

[4]Ibid   Page 184.
[5]Ibid.   Page 162.

ELNATHAN TABER.

Elnathan Taber, Simon Willard's best apprentice, was a native of Dartmouth, Mass., a son of Thomas and Elizabeth (Swift) Taber.[1] In the Roxbury Records is found the following entry:

"Elnathan Taber, son of Thomas and Elizabeth Taber, born at Dartmouth, in the county of Bristol, the 14th of February, 1768.[2]"

It was the custom in those days when a stranger arrived in town, to be recorded by the Town Authorities, which would account for the entry in the records. There is no evidence to show whether Taber's parents came with him or not. Drake[3] says, "Taber Street, originally named Union, laid out in 1802, was named after Elnathan Taber, a native of New Bedford, who came to Roxbury at the age of sixteen, served as an apprentice at Aaron Willard's, and was the first resident

[1]Extract from a letter from Mr. F. E. Smith, Roxbury, Mass., to the author.
[2]Rox. Rec. B. D. & M. 1630 to 1785, Manuscript Copy. Page 100.
[3]Town of Roxbury. R. C. Vol. 34. Page 207.

on the street." The author has been informed by a relative' of Taber's, that young Taber was nineteen when he arrived in Roxbury, but there is no documentary evidence to show which is correct. Taber evidently bought his place on Union Street about the time it was laid out.' Drake is greatly in error in saying Elnathan Taber was apprenticed to Aaron Willard. He was apprenticed to Simon Willard. The author has the authority for this, on the statement of his father, Z. A. Willard, who was well acquainted with Taber, who often told him that he was apprenticed to Simon Willard, and described his experiences there, and often heard Simon Willard himself say Elnathan Taber was his apprentice.

Taber was a life-long friend of Simon Willard and when Willard retired from business, bought most of his tools and the good-will of the business. There is no evidence to show just when Taber was apprenticed, but he set up in business for himself as soon as his time was out.

He married, January 8, 1797, Catherine Partridge.' She died Nov. 24, 1859, ae 85 years, 11 months 12 days.' He died Feb. 27, 1854, ae 86.'

Children of Elnathan and Catherine Taber.

Thomas, born, Roxbury, March 23, 1797.[7] Died, July 2, 1878.[8]

Catherine, born, Roxbury, April 7, 1802.[7]

Elizabeth Bourne, born, Roxbury, March 20, 1807.[7]

Abigail Haskell, born, Roxbury, Dec. 29, 1811.[7] Died, May 1, 1834.[6]

[4]Norfolk Deeds. Vol. 15. Page 168.
[5]Roxbury Marriages. 1632 to 1860.
[6]Roxbury Deaths. 1633 to 1860.
[7]Roxbury Births. 1632 to 1844.
[8]Boston Deaths for 1878. No. 3792.

His place on Taber St. was a very pretty one, consisting of a dwelling house (part of which was still standing in 1905) with a garden and cherry trees in front, and at one side of the house was a small workshop where he made his clocks. Taber made a most excellent clock, fully as good as Simon Willard. He made the tall Hall Clock, Half Clock, and the Timepiece. A very good example of his Timepiece is shown in Plate 35. The glass front is thought to have been painted by Charles Bullard, and the dial has the inscription: "Elnathan Taber. 1840."

Like Simon Willard, Elnathan Taber was a tireless worker, and worked literally up to the day of his death. An abstract from the inventory[9] of his property is given here as showing how few tools the old time clock-makers seemed to work with.

" Lot of old Files	$5.00	Lot of Hammers	$2.00	$7.00
Screw plates	5.00	Two Hand Lathers and saws	1.50	6.50
Old Brass	1.50	Sundry lots in brass	1.00	2.50
Time-piece case	.50	Iron stakes, squares, etc.	1.00	1.50
Chest and contents	.75	Three iron vises	3.75	4.50
One Lathe	.25	Cutting Engine	10.00	10.25
Depthing tool	5.00	Regulator movement	4.00	9.00
Large chest	1.25	Show case	3.50	4.75
Money scales and eye glass, thermometer	1.00	Time-piece	2.00	3.00
Six old clocks	6.00	Boxes $2.00, Old iron	1.00	9.00
Eight day brass clock				25.00

David P. Davis ⎫
Daniel Jackson ⎬ Appraisers."
Joseph Bugbee ⎭

[9]Norfolk Probate 17792.

PLATE 35

ELNATHAN TABER

F. E. SMITH
ROXBURY, MASS.

TIMEPIECES OWNED BY

WILLIAM CUMMENS

FRANCIS H. BIGELOW
CAMBRIDGE, MASS.

PLATE 36

TIMEPIECE
WILLIAM KING LEMIST
OWNED BY
HENRY CLAP KENDALL
DORCHESTER

GRAVITY CLOCK
PHILANDER J. WILLARD
OWNED BY
EDWIN K. JOHNSON
ASHBY, MASS.

The cutting engine mentioned in the inventory is the one Elnathan Taber bought of Simon Willard when he retired from business in 1839. It is given a particularly low valuation. After Simon Willard's retirement, Elnathan Taber made clocks for Simon Willard, Jr., & Son, from 1838 to 1854. His son, Thomas Taber, continued his father's business, but the author has never seen or heard of any clocks made by him, and the author's father states that Thomas Taber never made any clocks for Simon Willard, Jr., & Son.

WILLIAM CUMMENS.

William Cummens (also spelled in records as Cummins, Cummings, Cummengs, Comins), is supposed to have been a native of Roxbury, but the writer has been unable to find any definite record of his parents, or his birth, although perhaps he was the son of Jacob and Sarah (Bugbee) Cummens.[1] He was born about 1768. He was another of Simon Willard's apprentices, but just when apprenticed, the writer has been unable to ascertain, perhaps about the same time with Elnathan Taber. He lived on the "northwest corner of Taber and Winslow Streets,[2]" where he engaged in clock-making after serving his apprenticeship. He made all his clocks in a little room in his house, which was a quaint one-story building with an attic, not having a shop like Taber.

He married in Roxbury, March 10th, 1793, Polly Mayo.[3] She died in Roxbury, April 29, 1832, ae 69.[4] He died in Roxbury, April 20, 1834, ae 66.[4]

Children of William and Polly Cummens.

William, born, Roxbury, Sept. 15, 1793.[5]
Sally, born, Roxbury, July 1, 1795.[5] Died, Roxbury, Jan. 15, 1862.[6]
Mary, born, Roxbury, Nov. 10, 1798.[5] Died, Roxbury, Jan. 24, 1862.[6]

Very little is known about his son William. He was known to have assisted his father in his business, and his name appears as a witness in a deed executed by his father

[1] Rox. B. M. & D. 1630 to 1785. Manuscript Copy. Page 161.
[2] Drake, Town of Roxbury. R. C. Vol. 34. Page 207.
[3] Roxbury Marriages. 1632 to 1860.
[4] Roxbury Deaths. 1633 to 1860.
[5] Roxbury Births. 1632 to 1844.
[6] Roxbury Deaths. 1860 to 1867.

in 1816,[7] after which there is no further record of him, that
the author can find, and no mention of him is made in the
administration of William Cummens' estate; only the names
of the two daughters appear.[8] The inventory of his property
speaks of his house and land lying between Summer and
Union Streets, Roxbury. Among the inventory of his personal estate is the following:

> 29 Timepieces complete — $10.00 $290.
> Clock-makers tools and sundries in shop   85.
>
>   Appraisers.
>   Elijah Lewis.
>   Elnathan Taber.
>   Aaron D. Williams.

Not a very high valuation is given on the Timepieces
which would bring four times the amount to-day, and again
will be noticed the small value for the tools. It will be observed that Elnathan Taber, his near neighbor, is one of
the appraisers. William Cummens was the owner of Pew
No. 21, lower floor, Fifth Meeting House, Roxbury, in
1793.[9] The family is probably extinct.

William Cummens made a very good clock, making the
Hall Clock, Half Clock, and Timepiece. Good specimens of
his work are in existence, but are not often found. The
clock (Plate 35) is a good specimen of his Timepiece,
although the brass eagle on top is not the original ornament. It has very finely painted glass fronts in very bright
colors, which are thought to be Charles Bullard's work. The
tall Hall Clock, made in 1798 (Plate 34), is a very fine speci-

[7]Norfolk Deeds. Vol. 52. Page 23.
[8]Norfolk Probate No. 4785.
[9]History of First Church in Roxbury. W. E. Thwing. Page 202.

men of William Cummens' work. It is unusually tall for a clock of this kind, so much so as to be unique. The case, a very fine one, was made by Stephen Badlam (1751-1815), a cabinet-maker at Lower Mills, Dorchester. William Cummens generally had the inscription: "Warranted by William Cummens, Roxbury," on his clock dials.

WILLIAM KING LEMIST.

For a long time the author has thought that Elnathan Taber was the last of Simon Willard's apprentices it would be possible to identify and give an account of. Recently, however, inquiries were made of the author about a William Lemist supposed to have been a clock-maker in Roxbury, early in the nineteenth century. After an exhaustive inquiry the author has found sufficient evidence to justify him in including Lemist as one of Simon Willard's apprentices. William King Lemist, one of a family of twelve children, was born in Dorchester, April 18, 1791,[1] the son of John and Hannah (King) Lemist.[2] His brother, John Lemist,[3] lived nearly opposite Simon Willard,[4] and the two families were quite intimate, and there seems to be no doubt that William Lemist was apprenticed to Simon Willard, probably about 1806 or 1808. The author quotes a part of a letter in regard to this:[5]

"I have talked with my sister concerning the facts in your letter about William Lemist, she says that our Mother[6] always spoke of William Lemist as learning his trade of Simon Willard."

Lemist made a clock in 1812 as a wedding present for his sister, Hannah,[7] who married Henry Clap of Dorchester,

[1]R. C. Vol. 21. Page 206.
[2]R. C. Vol. 21. Page 237.
[3]R. C. Vol. 21. Page 198.
[4]R. C. Vol. 34. Page 154.
[5]Letter from Mr. Henry Clap Kendall, Dorchester.
[6]Hannah Lemist Clap, daughter of Henry and Hannah Clap. R. C. Vol. 21. Page 317.
   Married Volney Kendall. R. C. Vol. 36. Page 153.
[7]R. C. Vol. 21. Page 340.

the same year.   This clock, marked "William Lemist, 1812,"
is still in existence, and is highly valued by the family.

William Lemist did not continue in the business long
after serving his apprenticeship.

> "His health was poor and he went on a sea voyage for improvement.
> The name of the vessel we do not remember, but my mother did.   He was
> wrecked or died in the wreck of the ship off the coast of Africa, Oct., 1820,
> aged 29 years.   This is copied from family records in possession of my
> sister." [8]

William Lemist's life was so short that he evidently
made very few clocks. Apparently he made only the Timepiece (Plate 36).   There is no evidence he ever made the tall
Hall clock or the Half clock. This clock is in good condition and shows that Lemist followed very closely the
methods of his teacher, Simon Willard.   The method of lock-
ing the door of the clock, the fastening of the dial by little
hooks, and the shape of the hands are all copied from his
teacher.   The front of the clock has an enamelled and
gilded moulding.   The dial has the inscription: "Made by
William Lemist, 1812."   The glass paintings are somewhat
worn, but still show the original design.   The centrepiece
has a small medallion of a frigate under full sail.   The artist
who painted the glass is not known, but from its general
design, the author is inclined to think it was painted by
the artist who painted for the Aaron Willards.   Very curiously his brother John Lemist was lost in the burning of
the steamer Lexington, in Long Island Sound, Jan. 13, 1840. [9]

[8] Letter from Henry Clap Kendall, Dorchester.
[9] Rox. Rec. Deaths.   1633 to 1844.

CHARLES BULLARD.

Charles Bullard, a native of Dedham, Mass., son of William and Lydia (Whiting) Bullard,[1] was born August 13, 1794.[2] He married, November 10, 1822, Elizabeth Paul of Dedham.[3] She died May 21, 1884.[4] He died July 29, 1871.[5]

Children of Charles and Elizabeth Bullard.

> Ann Elizabeth, born, Roxbury, Nov. 5, 1823.[6]
> Charles Otis, born, Roxbury, Sept. 5, 1825.[6]
> Mary Lewis, born, Roxbury, April 30, 1827.[6]

Charles Bullard was the apprentice and successor of the English artist who painted the glass fronts and clock dials for Simon Willard. Just when he was apprenticed and where he worked while an apprentice, the writer has been unable to ascertain, but it was somewhere in Roxbury. After the expiration of his apprenticeship, Charles Bullard set up in business for himself. He had a shop at or in rear of No. 843 Washington St., Boston Neck.[7] He continued in this shop from 1816 to 1844,[7] advertising as an ornamental painter. He evidently had his residence in Roxbury, as in 1822 he is mentioned as being in Roxbury, and in 1824, he is taxed in the second parish of Roxbury for $1.00 poll,

[1]Dedham Records. B. M. & D. 1635 to 1845. Page 206.
[2]Ibid. Page 142.
[3]Ibid. Page 180.
[4]Dedham Records. Abstract of deaths. 1844 to 1890. Page 20. No. 69.
[5]Ibid. Page 20. No. 67.
[6]Roxbury Births. 1632 to 1844.
[7]Boston Directories. 1816 to 1844.

and $266.67, personal,[8] and the births of his children are all recorded in Roxbury.

At his shop on Washington St., Charles Bullard did work for all the surrounding clock-makers and doubtless for many of those out of town. After 1844 on the general breaking up of the clock-making colony on Boston Neck, Charles Bullard returned to Dedham, where he lived on Chestnut St.,[9] until his death.

Charles Bullard did very beautiful work, the decorating of his clock dials being especially fine. In his painting of glass fronts for Timepieces, although his work was very fine, he never quite approached the beauty of his English teacher's work. Like his teacher, Charles Bullard painted exclusive designs for Simon Willard's Timepieces (Plates 21 and 22). The writer has never seen them on any other clock-makers' timepieces. Charles Bullard painted glass fronts for the Aaron Willards, William Cummens, Elnathan Taber, and many others, but in painting for these he followed an entirely different line of designs. After Simon Willard's retirement in 1839, Charles Bullard was employed by Simon Willard, Jr., & Son to paint clock dials. He continued to do work for them after he returned to Dedham. His last work for them as recorded by an entry in their books, was Oct. 16, 1865, for decorating a clock dial. He left no successor in the business that the author can find, although it is probable he had apprentices.

[8]Roxbury Tax Lists, for 1824. Second Parish.
[9]Dedham Tax Lists. First Parish. 1870. Page 7.

WILLIAM FISK.

William Fisk, son of Samuel and Abigail (White) Fisk, was born in Watertown, Mass., Dec. 20, 1770.[1] He, with his elder brother, Samuel, born in Watertown, Sept. 24, 1769,[1] first appears in Roxbury, Mass., in 1792, giving their place of residence as Roxbury, and stating their occupation as cabinet-makers.[2] Some time after 1796, they moved to Boston, on the Neck,[3] where they had a shop next to Aaron Willard.

Samuel Fisk, who had married in 1794, Hannah Babcock of Milton,[4] died in 1797,[5] and William Fisk continued in the business at the same place until he died.[6] His place is described in 1798, as being "a lot of land, 30,000 square feet with a shop thereon 40 x 20, next to Aaron Willard,[7] and a lot of land with dwelling house, barn, etc., next to land of John Davis, Washington St., Boston Neck."[8] It was not until 1825[9] that houses began to be numbered to any extent on Boston Neck, when William Fisk is recorded as having his shop at No. 841 Washington St. and his dwelling house at No. 839. These numbers continued unchanged until his death.

William Fisk married, May 8, 1794, Eunice White of

[1] Bond Genealogical History of Watertown. 2nd Edition. Page 212, 213.
[2] Suffolk Deeds. Vol. 173. Page 261.
[3] Suffolk Deeds. Vol. 183. Page 242.
[4] R. C. Vol. 30. Page 464.
[5] Suffolk Probate. Vol. 95. Pp. 29, 42, 102
[6] Boston Directories. 1798 to 1844.
[7] R. C. Vol. 22. Page 107.
[8] Ibid. Page 410.
[9] Boston Directory for 1825.

Watertown.[10]   He died June 11, 1844.[11]   In his later years he seems to have changed his occupation, for in his will he calls himself a surveyor.[12]   He had a large family.[13]

William Fisk had a great reputation as a cabinet-maker. He made nearly all of Simon Willard's clock-cases from about 1800 to 1838, and also made them for the Aaron Willards and other clock-makers.   His work was very fine, although his specialty was furniture, especially inlaid work.   He had many apprentices.   The author has been unable to ascertain whom William Fisk and his brother were apprenticed to, unless perhaps it was to Nehemiah Munroe, a cabinet-maker who lived in Roxbury.[14]

[10]R. C.   Vol. 30.   Page 321.
[11]Boston Deaths.   1810 to 1848.
[12]Suffolk Probate.   Vol. 142.   Page 392.
[13]Bond Genealogical History of Watertown.   2nd Edition.   Page 213.
[14]Drake.   R. C.   Vol. 34.   Pp. 159.

CHARLES CRANE CREHORE.

Son of John Shephard Crehore and Hannah (Lyon) Crehore, was born at Dorchester, October 8, 1793.[1] He lived in Milton the early part of his life, and probably learned his trade of his father, who had a chair manufactory on the Lyman Davenport place.[2] He married, October 1, 1826, Chloe B. Hartwell, of Canton,[3] Mass. He was a very fine cabinet-maker. He made the cases for the clocks made by Simon Willard, Jr., and Benjamin F. Willard, (Plate 25), and also some for Simon Willard, Sr., and many other clock-makers. He moved to Boston about 1858, residing at 98 Charles St.[4] He died in Boston, February 12, 1879.[5] He left no successor to his business.

[1]Milton Records. B. M. & D. 1662 to 1848. Page 19.
[2]History of Milton. Page 178.
[3]Canton, Mass. B. M. & D. 1715 to 1845. Page 241.
[4]Boston Directory for 1858.
[5]Boston Deaths for 1879. No. 951.

THE WILLARD FAMILY
OF ASHBURNHAM AND ASHBY, MASS.,
CLOCK-MAKERS.

There seems to be very little known about this family, two of whom, Philander Jacob Willard and Alexander Tarbell Willard, brothers, carried on a prosperous business for over fifty years in Ashburnham and Ashby. They have often been confused with the Willards of Grafton, Mass. Stearns, in his History of Ashburnham, gives a brief account of them, but outside of this, the author was unable to find out much about them. Z. A. Willard says he never heard of them, and has never seen any of their clocks.

Desirous of getting more information about them, the author visited Ashby, and was fortunate enough to meet Mr. Edwin K. Johnson, a resident of that town, who was an intimate friend of the Willard brothers. Mr. Johnson gave the author much valuable information, and showed all the places of interest connected with them. This has enabled the author to give a fuller and more interesting history of this branch of clock-makers, than he had hoped for. They were sons[1] of Jacob[5] Willard (Henry[4] Henry[3] Henry[2] Simon[1]), and like the Willards of Grafton, were lineal descendants of Major Simon Willard. Jacob[5] Willard was born in Harvard, Mass., July 20, 1734, son of Henry[4] and Abigail (Fairbank) Willard.[2] He moved to Ashburnham, where he carried on the occupation of farming. He married

[1]History of Ashburnham, Mass. Ezra S. Stearns. Pages 980-981.
[2]History of Harvard, Mass. Henry S. Nourse. Page 573.

PLATE 37

OLD CLOCK FACTORY OF ALEXANDER T. WILLARD
ASHBY, MASS.

PLATE 38

MANSION HOUSE OF ALEXANDER T. WILLARD
ASHBY, MASS.

July 25, 1771,[3] Rhoda Randall of Stow, Mass. He died
Feb. 22, 1808.[4]

Children of Jacob and Rhoda (Randall) Willard.

> Philander Jacob, born, Sept. 29, 1772.[4]
> Alexander Tarbell, born, Nov. 4, 1774.[4]
> Ame, born, Dec. 18, 1777.[4]
> Katy, born May 24, 1781.[4]

The two brothers spent their boyhood on their father's
farm in Ashburnham, which was very near the Ashby line.
The old homestead was burned down some years ago. They
had a limited education, such as the district school furnished, and their mechanical faculties developed early.
They seem to have been intimate friends from the first with
a family named Edwards, who lived in Ashby. Two of
these, Abraham and Calvin Edwards, were gold- and silver-
smiths. Calvin, from 1789 to 1797,[5] when he died. Abraham
seems to have abandoned the goldsmith's business in 1794,[6]
and taken up clock-making, which he carried on up to
the time of his death in 1840.[7] The author was under the
impression that Philander J. and Alexander T. Willard
were apprentices of the Edwards, but Mr. Johnson says
they were not, they taught themselves. The author gives
the following extract from a letter from Mr. Johnson in
regard to Abraham Edwards:

> "Abraham Edwards once lived upon the place now owned (1901) by
> Mr. Frank W. Wright. When Mr. Edwards came to Ashby, he made pewter

[3]Ibid. Page 514.
[4]Ashburnham Vital Records.
[5]Middlesex Deeds. Vol. 100. Page 401. Vol. 104. Page 490. Vol. 112. Page 426.
[6]Middlesex Deeds. Vol. 115. Pages 475-479.
[7]Middlesex Probate. 23788.

buttons. Afterwards he (after taking an old clock to pieces) established clock-making in the village of Ashby. Upon the face of his clocks was a picture of his house with a horse-chestnut upon either side. West of his house was a shed, a clock-shop, a store, and a barn, all connected. When Mr. Edwards first came to Ashby he boarded with Lieu. Barrett.

This sketch of Mr. Edwards was told to the daughter (Mrs. L. Clementine Gates) of Mr. Jonas Barrett Damon, who worked for Mr. Edwards helping him to make the clocks.

The author is inclined to think that these Edwards might possibly be related to John Edwards (16..-1746),[s] the celebrated silversmith of Boston, Mass. It would be of interest to know if they were.

[s]Suffolk Probate. Vol. 38. Pages 514, 515.

PHILANDER JACOB WILLARD

Eldest son of Jacob[5] and Rhoda (Randall) Willard, was born in Ashburnham, Sept. 29, 1772.[1] His boyhood was spent on his father's farm. Just when he began clock-making is not known, but he made clocks in Ashburnham until 1825. In 1796, he married Rhoda Wheeler,[1] born in Ashby, Oct. 18, 1773. They were divorced[1] and he married, 2nd, Hannah Parker Snow of Dublin, N. H.[2] He died in Ashby, Dec. 26, 1840.[1]

Child of Philander Jacob and Hannah (Parker) Willard.

Augustine Horace, born May 18, 1809.[2]

This is the only child found recorded. In 1825, Philander J. Willard moved to Ashby, and was associated with his brother Alexander T. Willard in the clock-making business until his death in 1840. He did not have a separate house, but lived with his brother.

Philander J. Willard was a very fine workman. The author quotes a portion of a letter from Mr. Edwin K. Johnson in regard to him:

"I would say in my note to you I did not write very much of Philander Jacob Willard, but I did not intend to give the credit all to Alexander Tarbell Willard, for Philander Jacob Willard had not only the ingenuity, but that persistent faculty and patience, a worker always at his desk, and if I was asked which I called the greater I should say Jacob. He made my clock. I think he spent as many as five years on the clock."

This clock (Plate 36) is quite a curiosity and was called

---

[1]Stearns. History of Ashburnham. Pages 980-981.
[2]Ashburnham Vital Records.

by the maker a Perpetual Motion clock. It may be described as a gravity clock. The case is made of mahogany, about twenty-two inches long and six inches wide, and is in two pieces, a baseboard and the clock proper. The baseboard is fitted with a square steel peg, about two inches long, with a slot at one end. The clock proper has a dial six and one-half inches in diameter. The case enclosing the movement is of brass, and can be unscrewed to allow the works to be examined or cleaned, the whole being air-tight. At the back of the case about two-thirds of the way up is a slot about six inches long, exposing a steel rod with a hole cut to receive the slotted peg of the baseboard; the clock is thus suspended on the peg, pendulum-wise, being always perpendicular. The steel rod is ratcheted and engages cog wheels which actuate an escapement of some sort. The back of the case proper is cut out wherever possible, and weighted with lead, the whole case weighing about thirty pounds. In fact the clock is the weight. To start the clock it is hung on the steel peg on the baseboard and pushed up as far as the length of the slot in the back of the case will allow. The weight of the clock pulling down on the ratcheted rod sets the clock going. It takes about four days for the clock to fall the length of the slot, when it must be pushed up again.

The whole clock is a fine piece of workmanship, beautifully finished. The case is ornamented with mother-of-pearl buttons let into the wood, a brass top ornament, and brass side arms. It is difficult to understand, however, where the perpetual motion comes in. The author gives the above description of this clock with some diffidence as he was not allowed to take the clock to pieces to properly examine the movement, and also is not an expert in complicated clock movements.

The author also quotes a letter from the present owner, Mr. Edwin K. Johnson, in regard to this clock:

"It was made about one hundred years ago. Mr. Willard intended to make it a perpetual motion. He spent about five years on it. There is no pendulum, weights, springs, key or winding up to it. I do not think there is a man in Boston that could set it going or repair it. The clock ran for years in the old shop (Plate 37). When he died the old clock stopped. There was no one to start it. It was sent to a noted German clock-maker in New York. He never could make it go. Then it was sent to Boston. They could not start it. I tried to buy it for fifty years. I was in a clock repairer's shop one day, and I saw my old friend, the clock. The man stepped out. I went to the clock, and it started for me. The man came in, he said he could not start it, and he had worked on it for a great while, and I finally bought it for what he asked for repairing it."

The clock was not running when the author looked at it, and Mr. Johnson stated that it had not run for a number of years. This clock is another instance of how inventors incline to complicated mechanism.

ALEXANDER TARBELL WILLARD

Youngest son of Jacob[5] and Rhoda (Randall) Willard, born in Ashburnham, Nov. 4, 1774. His boyhood was very much the same as his brother's, Philander Jacob Willard's. He early developed the mechanical faculty, and certainly as early as 1796, was making clocks with his brother in Ashburnham. In 1800, he removed to Ashby, Mass., where he lived until his death. He married Tila Oakes of Cohasset. Marriage published, May 24, 1800.[1] She died April 17, 1860.[2] He died Dec. 4, 1850. They had a large family. His will mentions six children.

Children of Alexander[6] T. and Tila (Oakes) Willard.

Caroline Cutler.[2]

George A.[3]

Alexander T.[3]

Catherine Cushing.[3]

Lysander B.[3]

Charles H.[3]

Emma A.[3]

There are none of the family now resident in Ashby, all having either died or moved away. Alexander T. Willard became widely known as a manufacturer of clocks. He was postmaster of Ashby from 1812 to 1836. In 1805, he, associated with others, among them Abraham Edwards, incorporated and laid out a turnpike road, called the Ashby Turnpike.[4] This road ran from the State line between New

[1]Ashburnham Vital Records.
[2]Grave Stone. Old Ashby Cemetery.
[3]Middlesex Probate. 37384.
[4]Laws and Resolves of Massachusetts. 1805. Chapt. 30. Page 436.

Hampshire and Massachusetts through Ashby to Townsend Plains, Mass. It was a failure, and its failure carried down Alexander T. Willard and a dozen others. When he removed from Ashburnham to Ashby in 1800, he built the little factory (Plate 37) which is still standing (1909). Here he made all his clocks and other articles. The old dial of the clock that advertised his business is still to be seen on the building, on the right, close to the door. In 1809, he built the house (Plate 38) where he lived until his death. This house, largely built of brick, is in excellent preservation and was long the show place of the town, and is still considered the finest house there.

Alexander T. Willard and likewise his brother were very ingenious and skilful workmen, and did a large business, which was principally on orders. He made the tall Hall clock, Wooden clock, Musical clock, Church or Turret clock, and the Timepiece. With the exception of the clock made by Philander J. Willard (Plate 36), the author has never seen any of the clocks made by the Willard brothers, but their clocks were well made and were excellent time-keepers.[5] Whether their clocks were marked with their names, the author is unable to say. The inventory of the stock and tools in the shop of Alexander T. Willard[6] is given and is another illustration of how few tools the clock-makers of those times seemed to work with.

1 Lathe and Turning Tools	$2.50	Forge bellows and tools for	
3 Clock Engines and Measuring		casting	$2.00
Machine	1.00	Framing Tools	1.00

[5]Statement by Mr. Edwin K. Johnson.
[6]Middlesex Probate. 37384.

Watch Lathes and Tools	3.00	Clock-cases and parts of clocks	.25
Lot of files, shears, awls	1.00	1 Clock	1.80
3 Wooden clocks and cases	3.00	1 Wooden Timepiece	1.50
1 Brass do.	4.00	4 Clock cases	.50
1 Marble clock	6.00		

Alexander T. Willard was a true Yankee, and did not confine himself to clock-making exclusively. He also made the old-fashioned Theodolite or Compass, Gunters Chains, Scales, Timers (the old perambulator or odometer), Seraphines (the forerunner of the Reed Organ), Rifles (used to sharpen scythes), and repaired watches.[7] His market for clocks was western Massachusetts, Vermont, and New Hampshire. His clocks do not seem to have reached Boston and its vicinity, possibly because competition was too great to make it an object. None of his children carried on the business after his death.

[7]Information given by Mr. Edwin K. Johnson.

CHECK LIST OF SIMON WILLARD CLOCKS.

1. Turret Clock in First Church Meeting House at Dedham, Mass. Had an inscription on the iron frame, "Simon Willard, Roxbury, Mass." History. Given to the Church by Hon. Edward Dowse and Mrs. Shaw, and was set up March, 1820, when the meeting house was being remodelled. Now replaced by a modern clock.

Information from a letter to the author from Julius H. Tuttle, Dec. 16, 1907, also see Dedham Historical Register, Vol. 5, Page 3, for January, 1894.

2. Clock on Old State House, State St., Boston, Mass. Clock running and in perfect condition. Has the following inscription on a brass plate on the frame.

CHRONOMETER
Made by Simon Willard of Roxbury. Mass. in his 78th year for the City Hall Boston. Commenced Sept 1830 and finished Feb'y 1831

3. Clock for United States Senate Chamber, Washington, D. C. Made in 1801. Probably destroyed when the British troops burned the Capitol in 1814. See letter (Page 18).

4. Turret Clock for Jefferson College, Virginia. Made from plans and specifications by Thomas Jefferson, 1826. Destroyed by fire in 1895. Alumni Bulletin of the University of Virginia, Charlottesville, Va. Pages 111, 112, 113.

5. Turret Clock for First Church of Roxbury, Mass. Made in 1806 for $858. Replaced by a modern clock.

6. Large Gallery Clock for the inside of same Church. Made in 1804. Has original glasses by the English artist, full gilded (Plate 23). History of First Church of Roxbury. W. E. Thwing.

7. Turret Clock in Park St. Church, Boston, Mass. Made in 18—. Replaced in 1906 by modern clock.

8. Turret Clock in North Church, Newburyport, Mass. Set up in 1785. Destroyed by fire in 1861. J. J. Currier, History of Newburyport. Pages 160, 280.

9. Turret Clock on Boylston Market, Boston, Mass. Clock given by Mr. Ward Nicholas Boylston. Snow's History of Boston, page 332. When the Boylston Market was torn down the tower and clock were purchased and placed on Van Nostrand's Brewery in Charlestown, Mass. Still running and in good condition. From information given by the late Mr. Patrick Greene, of Boston.

10. Turret Clock in North Church Meeting House, Portland, Maine. Set up in 1802. Replaced by a modern clock in 1893. Letter of Simon Willard, dated Jan. 13, 1802

(Page 19).   Also letter of Simon Willard dated July 6, 1802 (Plate 7).

11.   Clock in Chief Clerk's office, Supreme Court, Washington.   Ordered by Associate Chief Justice Story.   Set up in 1837.   One of the last made by Simon Willard before he retired.   (See Plate 12.)

12.   Franzoni Clock.   Statuary Hall, United States Capitol, Washington, D. C.   Works made especially for the case.   Probably put in at the same time that Simon Willard put up the other clock in Chief Clerk's office.   Has the stamp, "Simon Willard & Son" (Plates 13 and 14).

13.   Turret Clock in First Parish Unitarian Church, Cambridge, Mass.   Made in 1832.   In fine condition and still running.   Has a brass disk engraved and enclosed in glass.

14.   Large Regulator Clock in office of the Massachusetts Hospital and Life Insurance Co., 50 State St., Boston, Mass.   Has solid mahogany case, had name "Simon Willard" on the dial, but name was painted out some years ago.   Dial made of heavy brass, originally had base piece and acorn top ornament.   A remarkably fine specimen of Simon Willard's work.   Information given by Wm. Bond & Co.

15. Large Regulator Clock in the basement of the Provident Savings Bank, Temple Place, Boston, Mass. Has original glasses in fine condition; date not known.

16. Dial Clock, Roxbury St., Roxbury, Mass. Large double dial clock put up on Mr. Child's house, Simon Willard's next door neighbor. Set up about 1780. Remained there upwards of 80 years. Now in possession of some clock collector. Letter from Mr. Benjamin James, Roxbury, Mass.

17. Harvard University, Cambridge, Mass. Large Regulator Clock in Room 4, University Hall. Made in 1829 (Plate 11).

18. Gallery Clock in Gore Hall. Has a modern oak case.

19. Gallery Clock in Divinity Hall. Presented to the college by a friend.

20. Large Double Dial Clock. Made for the old Boylston Bank, Boston, Mass. Was recently sold at auction. Present owner unknown.

21. Turret Clock on First Congregational Church, Falmouth, Mass. Has inscription, " B. F. Willard, 1840." Probably made by Simon Willard and put up by his son, Benjamin F. Willard.

22. Gallery Clock in same church. Has front glass painted by the English artist. Inscription on front glass, " S. Willard, Patent."

23. Gallery Clock in Second Church, Codman Square, Dorchester, Mass. Has inscription on the dial, " Presented by the Hon. James Bowdoin." Given in 1808. Almost identical with the clock in the First Church, Roxbury, Mass.

24. Large Turret Clock made for some church or public building in New York City. Author has been unable to locate it. (See No. 7, Page 131.)

25. Large Double Dial Clock, formerly stood in front of the old Cambridge Bank, Cambridgeport, Mass. Was bought by Mr. Gardiner M. Lane and is now in front of his stable at Manchester, Mass. It is still an excellent time-keeper.